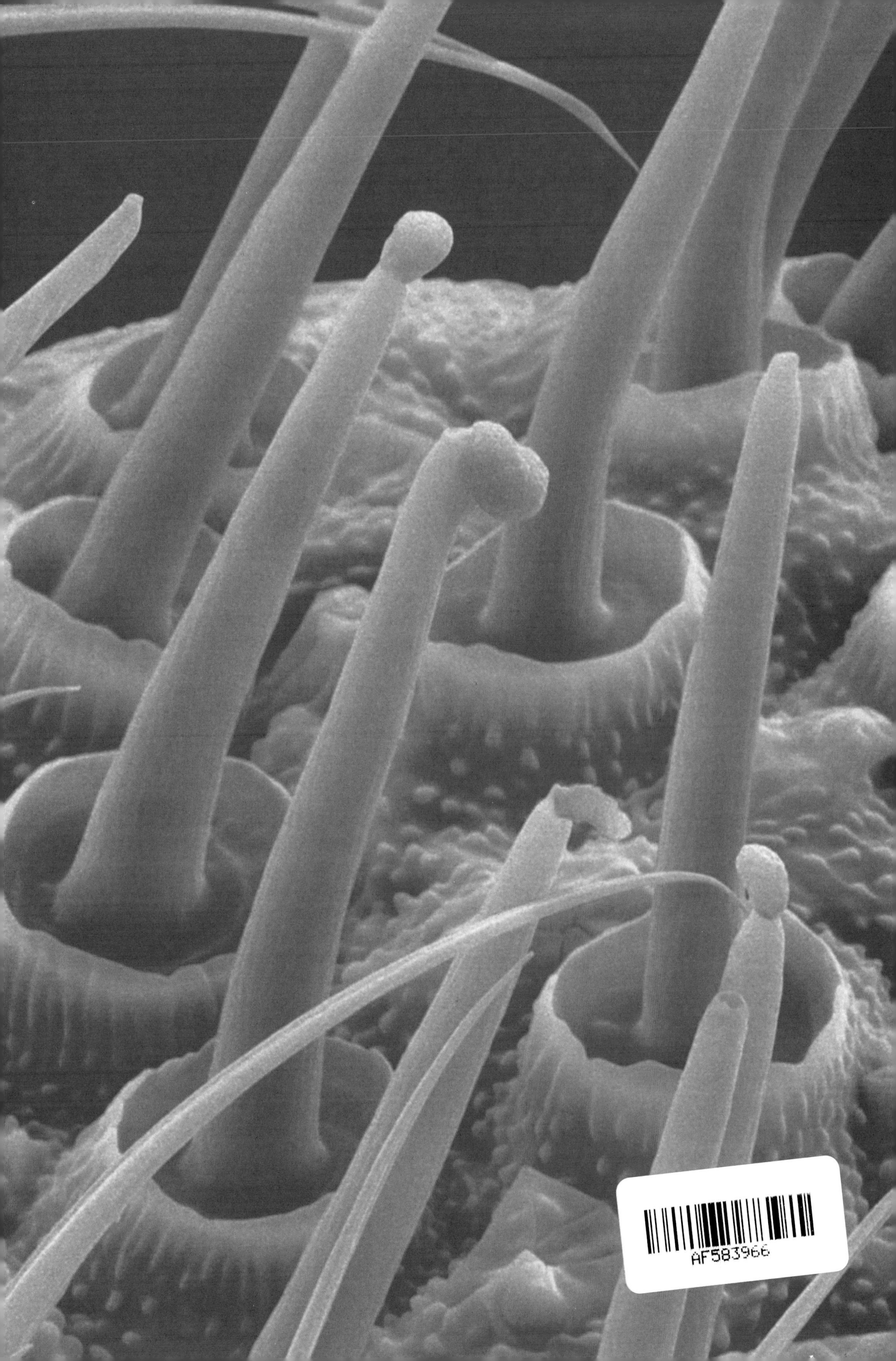

Pearson Australia
(a division of Pearson Australia Group Pty Ltd)
459–471 Church Street, Level 1, Building B,
Richmond, Victoria 3121
PO Box 23360, Melbourne, Victoria 8012

www.pearson.com.au

First published 2018 by Pearson Australia
2025 2024 2023 2022
10 9 8 7 6 5 4 3 2 1

Lead Publisher: Misal Belvedere
Project Manager: Michelle Thomas
Production Manager: Casey McGrath
Lead Development Editor: Fiona Cooke
Development Editors: Amy Sparkes and Haeyean Lee
Editor: Sally Woollett
Designer: Anne Donald
Rights and Permissions Editors: Samantha Russell-Tulip and Jenny Jones
Senior Publishing Services Analyst: Rob Curulli
Proofreader: Right Style Editing
Illustrator: DiacriTech
Printed in Malaysia (CTP-VVP)
ISBN 978 1 4886 1933 5

Pearson Australia Group Pty Ltd ABN 40 004 245 943

Acknowledgements
The following abbreviations are used in this list: t = top, b = bottom, l = left, r = right, c = centre.

Fotolia: Cog-Shared, p. 14t; Mathier, p. 14b.

Science Photo Library: Andrew Lambert Photography, p. 113; Carlos Clarivan, p. 17; Dennis Kunkel Microscopy, p. i; Mikkel Juul Jensen, p. 3.

Shutterstock: Anette Andersen, pp. 97-8; fotohunter, pp. 1-2; Nik Merkulov, pp. 50–1; Monticello, p. 21; Peter J. Wilson, pp. 150–1.

Contents

Contents

Module 3: Reactive chemistry

Module 4: Drivers of reactions

How to use this book

The *Pearson Chemistry 11 New South Wales Skills and Assessment* book provides an intuitive, self-paced approach to science education that ensures every student has opportunities to practise, apply and extend their learning through a range of supportive and challenging activities. While offering opportunities for reinforcement of key concepts, knowledge and skills, these activities enable flexibility in the approach to teaching and learning.

Explicit scaffolding makes learning objectives clear, and there are regular opportunities for student reflection and self-evaluation at the end of individual activities throughout the book. Students are also guided in self-reflection at the end of each module. In addition, there are rich opportunities to take the content further with the explicit coverage of Working scientifically skills and key knowledge in the depth studies.

This resource has been written to the new New South Wales Chemistry Stage 6 Syllabus and addresses the first four modules of the syllabus. Each module consists of five main sections:

- key knowledge
- worksheets
- practical activities
- depth study/ies
- module review questions.

Explore how to use this book below.

Chemistry toolkit

The Chemistry toolkit supports development of the skills and techniques needed to undertake practical investigations, secondary-sourced investigations and depth studies, and covers examination techniques and study skills. It also includes checklists, models, exemplars and scaffolded steps. The toolkit can serve as a reference tool to be consulted as needed.

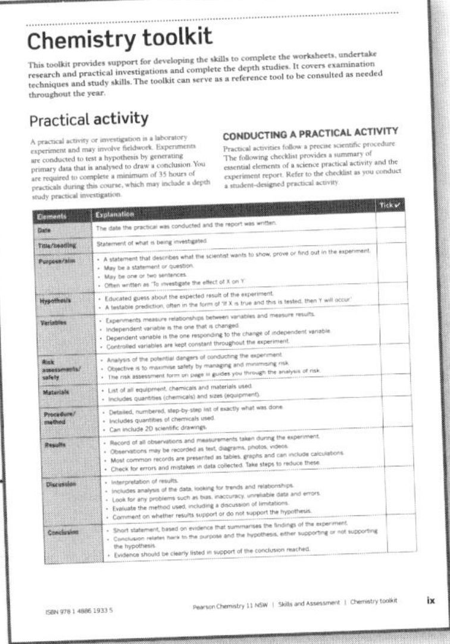

Chemistry toolkit

This toolkit provides support for developing the skills to complete the worksheets, undertake research and practical investigations and complete the depth studies. It covers examination techniques and study skills. The toolkit can serve as a reference tool to be consulted as needed throughout the year.

Practical activity

CONDUCTING A PRACTICAL ACTIVITY

MODULE 3 Reactive chemistry

Outcomes

Content

CHEMICAL REACTIONS

INQUIRY QUESTION What are the products of a chemical reaction?

PREDICTING REACTIONS OF METALS

INQUIRY QUESTION How is the reactivity of various metals predicted?

Module opener

Each book is split into the four modules of the syllabus, with the module opener linking the module content to the syllabus.

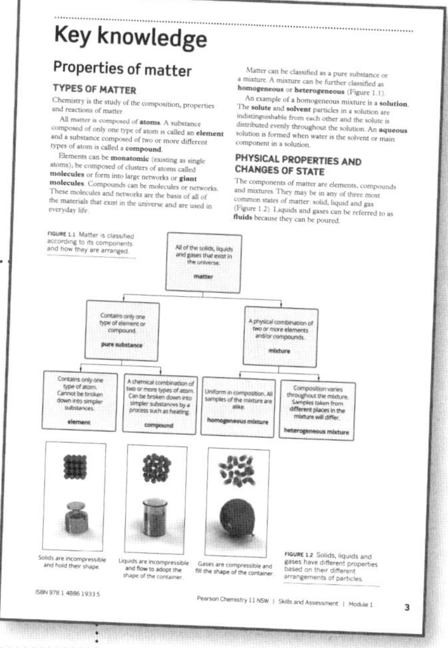

Key knowledge

Properties of matter

TYPES OF MATTER

PHYSICAL PROPERTIES AND CHANGES OF STATE

Key knowledge

Each module begins with a key knowledge section. This consists of a set of succinct summary notes that cover the key knowledge set out in each module of the syllabus. This section is highly illustrative and written in a straightforward style to assist students of all reading abilities. Key terms are bolded for ease of navigation. It also provides a ready reference for completing the worksheets and practical activities.

Worksheets

A diverse offering of instructive and self-contained worksheets is included in each module. Common to all modules is the initial 'Knowledge review' worksheet to activate prior knowledge, a 'Literacy review' worksheet to explicitly build understanding and application of scientific terminology, and finally a 'Thinking about my learning' worksheet, which provides a reflection and self-assessment opportunity for students. Each additional worksheet provides opportunities to revise, consolidate and further student understanding.

These worksheets function as formative assessment and are clearly aligned to the syllabus. A range of questions building from foundation to challenging are included within worksheets.

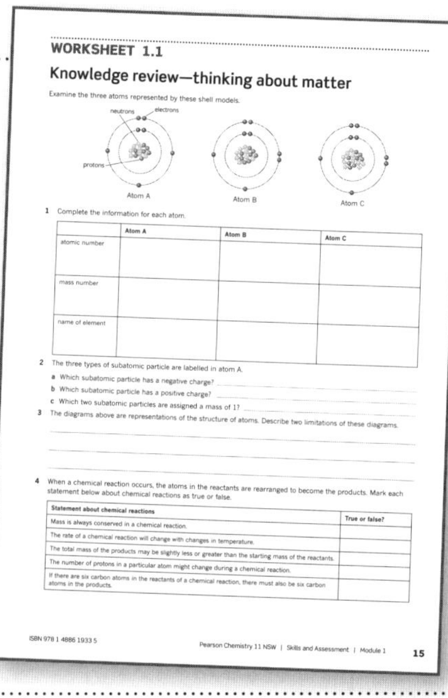

WORKSHEET 1.1

Knowledge review—thinking about matter

Examine the three atoms represented by these shell models.

1 Complete the information for each atom.

	Atom A	Atom B	Atom C
atomic number			
mass number			
name of element			

2 The three types of subatomic particle are labelled in atom A.

a Which subatomic particle has a negative charge?

b Which subatomic particle has a positive charge?

c Which two subatomic particles are assigned a mass of 1?

3 The diagrams above are representations of the structure of atoms. Describe two limitations of these diagrams.

4 When a chemical reaction occurs, the atoms in the reactants are rearranged to become the products. Mark each statement below about chemical reactions as true or false.

Statement about chemical reactions	True or false?
Mass is always conserved in a chemical reaction.	
The rate of a chemical reaction will change with changes in temperature.	
The total mass of the products may be slightly less or greater than the starting mass of the reactants.	
The number of protons in a particular atom might change during a chemical reaction.	
If there are six carbon atoms in the reactants of a chemical reaction, there must also be six carbon atoms in the products.	

ISBN 978 1 4886 1933 5 Pearson Chemistry 11 NSW | Skills and Assessment | Module 1 15

PRACTICAL ACTIVITY 1.1

Separation techniques—purification of polluted water

Suggested duration: 50 minutes

INTRODUCTION

Water is never found pure in nature. It dissolves many impurities, and insoluble substances may remain in suspension. Any purification process needs to take account of the different types of substances found in water. Common water purification methods include:

- filtration: Insoluble particles are separated from water. In the laboratory, filtration can be achieved using different grades of filter paper.
- charcoal adsorption: Charcoal has the ability to adsorb many substances, that is, to hold them to its surface. Charcoal can be used to adsorb the dyes that colour water, or to adsorb substances that give water a foul odour or taste.
- distillation: Water is separated from other liquids and solids dissolved in it.
- oil–water separation: When allowed to stand undisturbed, a mixture of oil and water forms two layers, with the oil on top. The water can then be drained using a separating funnel. This method is based on the principle that oil and water have different densities and polarities, and are essentially insoluble in one another.

MATERIALS

- 100 mL polluted water
- activated charcoal
- separating funnel
- filter papers
- distillation equipment
- conductivity kit
- safety glasses

PURPOSE

- To devise and carry out a method for purification of polluted water.
- To obtain as large a volume, and as pure a sample, of water as possible from a sample of polluted water.

PRE-LAB SAFETY INFORMATION

Material used	Hazard	Control
conductivity kit	electricity	Construct the circuit with the power off. Do not touch the circuit while the power is on. Do not allow electrical equipment to come into contact with liquids.
polluted water	harmful substances in the water	Do not taste your polluted or 'purified water' sample. Wash your hands after handling polluted water.

Please indicate that you have understood the information in the safety table.

Name (print):

I understand the safety information (signature):

PROCEDURE

Record all of your procedure results in Table 1.

1 Record the volume, appearance and odour of your sample of water.

2 Using a voltage of 8 V, construct a simple circuit to test the electrical conductivity of the purified water. Be very careful not to let the electrodes touch each other.

3 Design a method of purifying your water sample. List the steps as a flow chart.

ISBN 978 1 4886 1933 5 Pearson Chemistry 11 NSW | Skills and Assessment | Module 1 27

Practical activities

Practical activities provide the opportunity to complete practical work related to the various themes covered in the syllabus. All practical activities referenced in outcomes within the syllabus have been covered. Across the suite of practical activities provided, students are exposed to opportunities where they design, conduct, evaluate, gather and analyse data, appropriately record results and prepare evidence-based conclusions directly into the scaffolded practical activities. Students also have opportunities to evaluate safety and any potential hazards.

Each practical activity includes a suggested duration. Along with the depth studies, the practical activities meet the 35 hours of practical work mandated at Year 11 in the syllabus. Where there is key knowledge that will support the completion of a practical activity, students are referred back to it.

Like the worksheets, the practical activities include a range of questions building from foundation to challenging.

Depth studies

Each module contains at least one suggested depth study. The depth studies allow further development of one or more concepts found within or inspired by the syllabus. They allow students to acquire a depth of understanding and take responsibility for their own learning. They also promote differentiation and engagement.

Each depth study allows for the demonstration of a range of Working scientifically skills, with all depth studies assessing the Working scientifically outcomes of Questioning and predicting and Communicating. A minimum of two additional Working scientifically skills and at least one Knowledge and understanding outcome are also assessed.

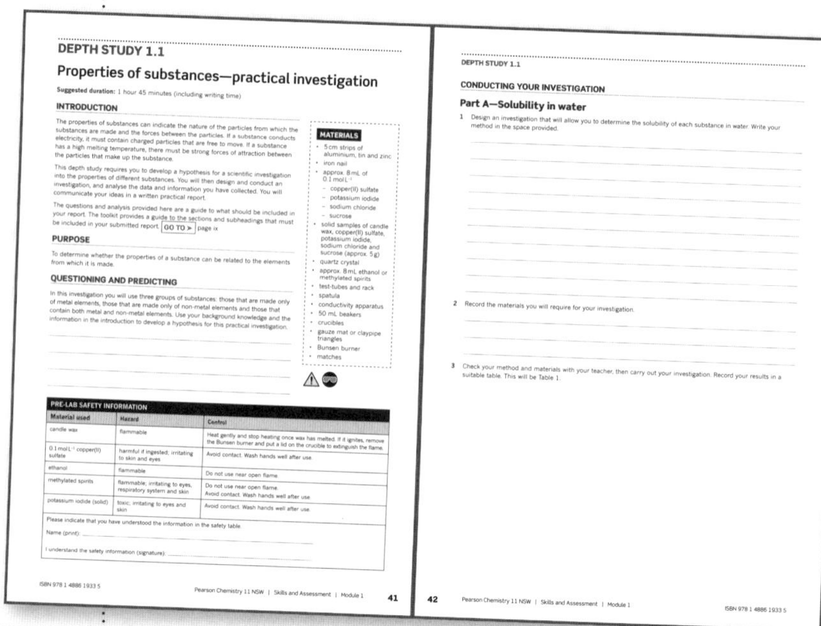

DEPTH STUDY 1.1

Properties of substances—practical investigation

Suggested duration: 1 hour 45 minutes (including writing time)

INTRODUCTION

The properties of substances can indicate the nature of the particles from which the substances are made and the forces between the particles. If a substance conducts electricity, it must contain charged particles that are free to move. If a substance has a high melting temperature, there must be strong forces of attraction between the particles that make up the substance.

This depth study requires you to develop a hypothesis for a scientific investigation into the properties of different substances. You will then design and conduct an investigation, and analyse the data and information you have collected. You will communicate your ideas in a written practical report.

The questions and analysis provided here are a guide to what should be included in your report. The toolkit provides a guide to the sections and subheadings that must be included in your submitted report. GO TO ➤ page ix

MATERIALS

- 5 cm strips of aluminium, tin and zinc
- iron nail
- approx. 8 mL of 0.1 mol L^{-1}
  - copper(II) sulfate
  - potassium iodide
  - sodium chloride
  - sucrose
- solid samples of candle wax, copper(II) sulfate, potassium iodide, sodium chloride and sucrose (approx. 5 g)
- quartz crystal
- approx. 8 mL ethanol or methylated spirits
- test-tubes and rack
- spatula
- conductivity apparatus
- 50 mL beakers
- crucibles
- gauze mat or claypipe triangles
- Bunsen burner
- matches

PURPOSE

To determine whether the properties of a substance can be related to the elements from which it is made.

QUESTIONING AND PREDICTING

In this investigation you will use three groups of substances: those that are made only of metal elements, those that are made only of non-metal elements and those that contain both metal and non-metal elements. Use your background knowledge and the information in the introduction to develop a hypothesis for this practical investigation.

PRE-LAB SAFETY INFORMATION

Material used	Hazard	Control
candle wax	flammable	Heat gently and stop heating once wax has melted. If it ignites, remove the Bunsen burner and put a lid on the crucible to extinguish the flame.
0.1 mol L^{-1} copper(II) sulfate	harmful if ingested; irritating to skin and eyes	Avoid contact. Wash hands well after use.
ethanol	flammable	Do not use near open flame.
methylated spirits	flammable; irritating to eyes, respiratory system and skin	Do not use near open flame. Avoid contact. Wash hands well after use.
potassium iodide (solid)	toxic; irritating to eyes and skin	Avoid contact. Wash hands well after use.

Please indicate that you have understood the information in the safety table.

Name (print):

I understand the safety information (signature):

ISBN 978 1 4886 1933 5 Pearson Chemistry 11 NSW | Skills and Assessment | Module 1 41

DEPTH STUDY 1.1

CONDUCTING YOUR INVESTIGATION

Part A—Solubility in water

1 Design an investigation that will allow you to determine the solubility of each substance in water. Write your method in the space provided.

2 Record the materials you will require for your investigation.

3 Check your method and materials with your teacher, then carry out your investigation. Record your results in a suitable table. This will be Table 1.

42 Pearson Chemistry 11 NSW | Skills and Assessment | Module 1 ISBN 978 1 4886 1933 5

ISBN 978 1 4886 1933 5

Module review questions

Each module finishes with a comprehensive set of questions, consisting of multiple choice, short answer and extended response, that helps students to draw together their knowledge and understanding and apply it to these styles of questions.

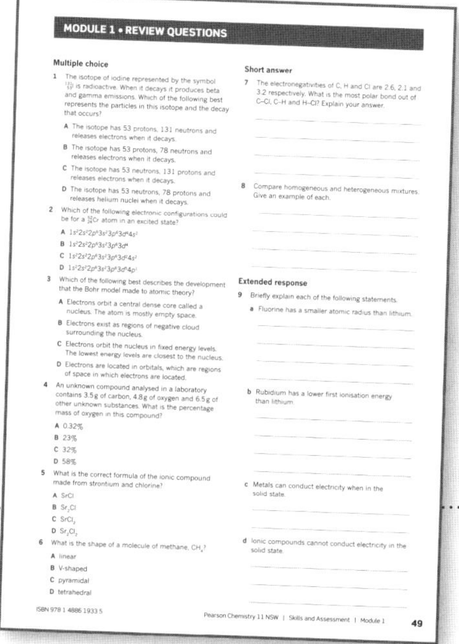

MODULE 1 • REVIEW QUESTIONS

Multiple choice

1 The isotope of iodine represented by the symbol $^{131}_{53}I$ is radioactive. When it decays it produces beta and gamma emissions. Which of the following best represents the particles in this isotope and the decay that occurs?

A The isotope has 53 protons, 131 neutrons and releases electrons when it decays.

B The isotope has 53 protons, 78 neutrons and releases electrons when it decays.

C The isotope has 53 neutrons, 131 protons and releases electrons when it decays.

D The isotope has 53 neutrons, 78 protons and releases helium nuclei when it decays.

2 Which of the following electronic configurations could be for a $^{52}_{24}Cr$ atom in an excited state?

A $1s^22s^22p^63s^23p^63d^44s^2$

B $1s^22s^22p^63s^23p^63d^4$

C $1s^22s^22p^63s^23p^63d^24s^2$

D $1s^22s^22p^63s^23p^63d^54p^1$

3 Which of the following best describes the development that the Bohr model made to atomic theory?

A Electrons orbit a central dense core called a nucleus. The atom is mostly empty space.

B Electrons exist as regions of negative cloud surrounding the nucleus.

C Electrons orbit the nucleus in fixed energy levels. The lowest energy levels are closest to the nucleus.

D Electrons are located in orbitals, which are regions of space in which electrons are located.

4 An unknown compound analysed in a laboratory contains 3.5 g of carbon, 4.8 g of oxygen and 6.5 g of other unknown substances. What is the percentage mass of oxygen in this compound?

A 0.32%

B 23%

C 32%

D 58%

5 What is the correct formula of the ionic compound made from strontium and chlorine?

A SrCl

B Sr_2Cl

C $SrCl_2$

D Sr_2Cl_2

6 What is the shape of a molecule of methane, CH_4?

A linear

B V-shaped

C pyramidal

D tetrahedral

Short answer

7 The electronegativities of C, H and Cl are 2.6, 2.1 and 3.2 respectively. What is the most polar bond out of C–Cl, C–H and H–Cl? Explain your answer.

8 Compare homogeneous and heterogeneous mixtures. Give an example of each.

Extended response

9 Briefly explain each of the following statements.

a Fluorine has a smaller atomic radius than lithium.

b Rubidium has a lower first ionisation energy than lithium.

c Metals can conduct electricity when in the solid state.

d Ionic compounds cannot conduct electricity in the solid state.

ISBN 978 1 4886 1933 5 Pearson Chemistry 11 NSW | Skills and Assessment | Module 1 49

Icons and features

The New South Wales Stage 6 Syllabus Learning across the curriculum content is addressed and identified using the following icons:

GoTo icons are used to make important links to relevant content within the book.

The **safety icon** highlights significant hazards indicating caution is needed.

The **safety glasses icon** highlights that protective eyewear is to be worn during the practical activity.

A **pre-lab safety box** is included. Students are to sign agreeing that they have understood the hazards associated with the material(s) in use and the control measures to be taken.

PRE-LAB SAFETY INFORMATION		
Material used	**Hazard**	**Control**
Conductivity kit	Electricity	Construct the circuit with the power off. Do not touch the circuit while the power is on. Do not allow electrical equipment to come into contact with liquids.
Polluted water	harmful substances in the water	
Do not taste your polluted or 'purified water' sample. Wash hands after handling polluted water.		
Please indicate that you have understood the information in the safety table. Name (print): ______ I understand the safety information (signature): ______		

Rating my learning

Rating my learning is an innovative tool that appears at the bottom of the final page of most worksheets and all practical activities. It provides students with the opportunity for self-reflection and self-assessment. It encourages students to look ahead to how they can continue to improve, and assists in highlighting focus areas for further skill and knowledge development.

The teacher may choose to use student responses to the 'Rating my learning' feature as a formative assessment tool. At a glance, teachers can assess which topics and which students need intervention for improvement.

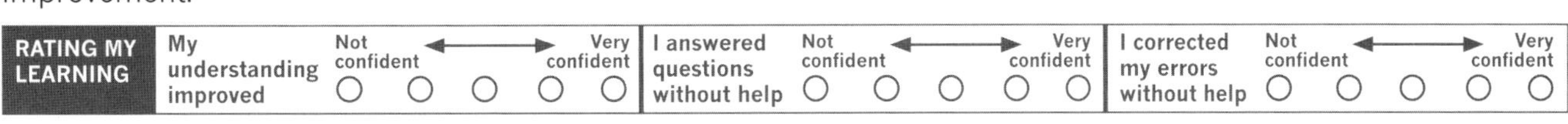

Teacher support

Comprehensive answers and fully worked solutions for all worksheets, practical activities, depth studies and module review questions are provided in the *Pearson Chemistry 11 New South Wales* Teacher Support. An editable suggested assessment rubric for depth studies is also provided.

Pearson Chemistry 11 New South Wales

Student Book

Pearson Chemistry 11 New South Wales has been written to fully align with the 2018 New South Wales Chemistry Stage 6 Syllabus. The Student Book includes the very latest developments and applications of chemistry and incorporates best practice literacy and instructional design to ensure the content and concepts are fully accessible to all students.

Skills and Assessment Book

Pearson Chemistry 11 New South Wales Skills and Assessment gives students the edge in preparing for all forms of assessment. Key features include a toolkit, key knowledge summaries, worksheets, practical activities, suggested depth studies and module review questions. It provides guidance, assessment practice and opportunities to develop key skills.

Reader+ the next generation eBook

Pearson Reader+ lets you use your *Student Book* online or offline on any device. Pearson Reader+ retains the look and integrity of the printed book. Practical activities, interactives and videos are available on Pearson Reader+ along with the fully worked solutions to the Student Book questions.

Teacher Support

The Teacher Support includes syllabus grids and a scope and sequence plan to support teachers with programming. It also provides the fully worked solutions and answers to all *Student Book* and *Skills and Assessment Book* questions, including all worksheets, practical activities, depth studies and module review questions. Teacher notes, safety notes, risk assessments and a laboratory technician's checklist and recipes are available for all practical activities. Depth studies are supported with suggested assessment rubrics and exemplar answers.

Pearson Digital
Access your digital resources at **pearsonplaces.com.au**
Browse and buy at **pearson.com.au**

ISBN 978 1 4886 1933 5

Chemistry toolkit

This toolkit provides support for developing the skills to complete the worksheets, undertake research and practical investigations and complete the depth studies. It covers examination techniques and study skills. The toolkit can serve as a reference tool to be consulted as needed throughout the year.

Practical investigation

A practical investigation or activity is a laboratory experiment and may involve fieldwork. Experiments are conducted to test a hypothesis by generating primary data that is analysed to draw a conclusion. You are required to complete a minimum of 35 hours of practical work during this course, which may include a depth study practical investigation.

CONDUCTING A PRACTICAL INVESTIGATION

Practical investigations follow a precise scientific procedure. The following checklist provides a summary of essential elements of a science practical investigation and the experiment report. Refer to the checklist as you conduct a student-designed practical investigation.

Elements	Explanation	Tick ✔
Date	The date the practical was conducted and the report was written.	
Title/heading	Statement of what is being investigated.	
Purpose/aim	• Statement that describes what the scientist wants to show, prove or find out in the experiment. • May be a statement or question. • May be one or two sentences. • Often written as 'To investigate the effect of X on Y.'	
Hypothesis	• Educated guess about the expected result of the experiment. • Testable prediction, often in the form of 'If X is true and this is tested, then Y will occur.'	
Variables	• Experiments measure relationships between variables and measure results. • Independent variable is the one that is changed. • Dependent variable is the one responding to the change in the independent variable. • Controlled variables are kept constant throughout the experiment.	
Risk assessments/ safety	• Analysis of the potential dangers of conducting the experiment. • Objective is to maximise safety by managing and minimising risk. • The risk assessment form on page xi guides you through the analysis of risk.	
Materials	• List of all equipment, chemicals and materials used. • Includes quantities (chemicals) and sizes (equipment).	
Procedure/ method	• Detailed, numbered, step-by-step list of exactly what was done. • Includes quantities of chemicals used. • Can include two-dimensional scientific drawings.	
Results	• Record of all observations and measurements taken during the experiment. • Observations may be recorded as text, diagrams, photos, videos. • Most common records are presented as tables, graphs and can include calculations. • Check for errors and mistakes in data collected. Take steps to reduce these.	
Discussion	• Interpretation of results. • Includes analysis of the data, looking for trends and relationships. • Look for any problems such as bias, inaccuracy, unreliable data and errors. • Evaluate the procedure used, including a discussion of limitations. • Comment on whether results support or do not support the hypothesis.	
Conclusion	• Short statement based on evidence that summarises the findings of the experiment. • Conclusion relates back to the purpose and the hypothesis, either supporting or not supporting the hypothesis. • Evidence should be clearly listed in support of the conclusion reached.	

RISK ASSESSMENT FORM

Five levels of safety should be considered in an investigation. The inverted pyramid ranks these levels in order of importance. The school and your teacher are responsible for reducing most of these risks. You as a student scientist can take measures to reduce the risks shown at the bottom of the hierarchy.

Complete a risk assessment form for each activity to identify possible risks for which you can take responsibility and to think of ways you can reduce risks to create a safe experiment environment.

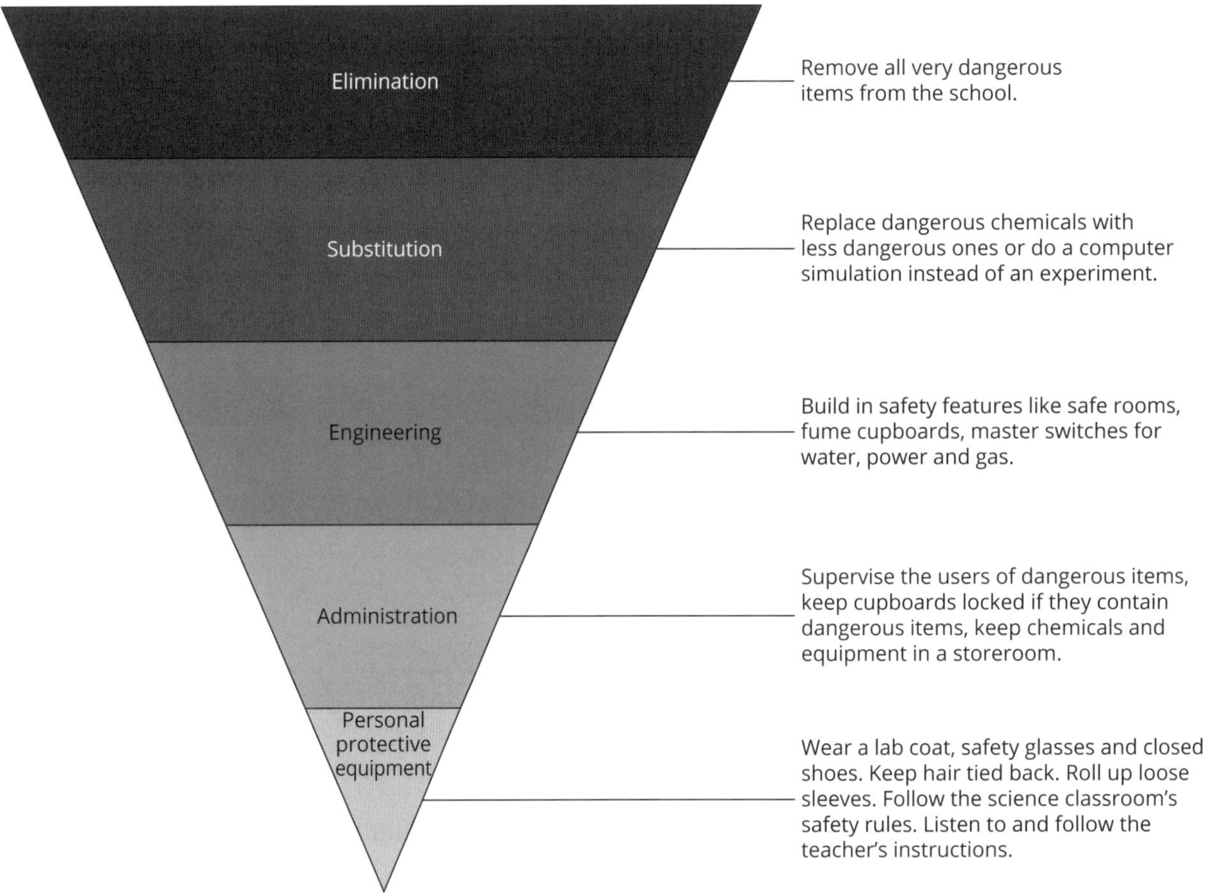

 ISBN 978 1 4886 1933 5

Risk assessment form		
What activity are you doing?		
Title or description of the investigation:		
List ...	**Identify any risks**	**State how you will ...**
equipment you will be using:		safely use each piece of equipment:
chemicals you will be using:		carefully use each chemical: carefully dispose of the chemicals:
ethical issues you need to consider:		ethically use animals in the laboratory: ethically use human participants in the investigation:
outdoor or fieldwork activities:		reduce any risks:
any other possible risks:		reduce these risks:

EXAMPLES OF PRACTICAL REPORTS

It can be difficult to gauge whether you have attained a high standard in your completed practical activity report. Looking at sample practical reports can help you identify what is required. Two sample practical reports are provided: one is prepared to a high standard while the second has room for improvement. The annotations draw your attention to key points to note on each practical report. These points are also reflected in the checklist, so you are able to use this as a tool to evaluate whether all requirements of the practical are complete.

High standard practical report

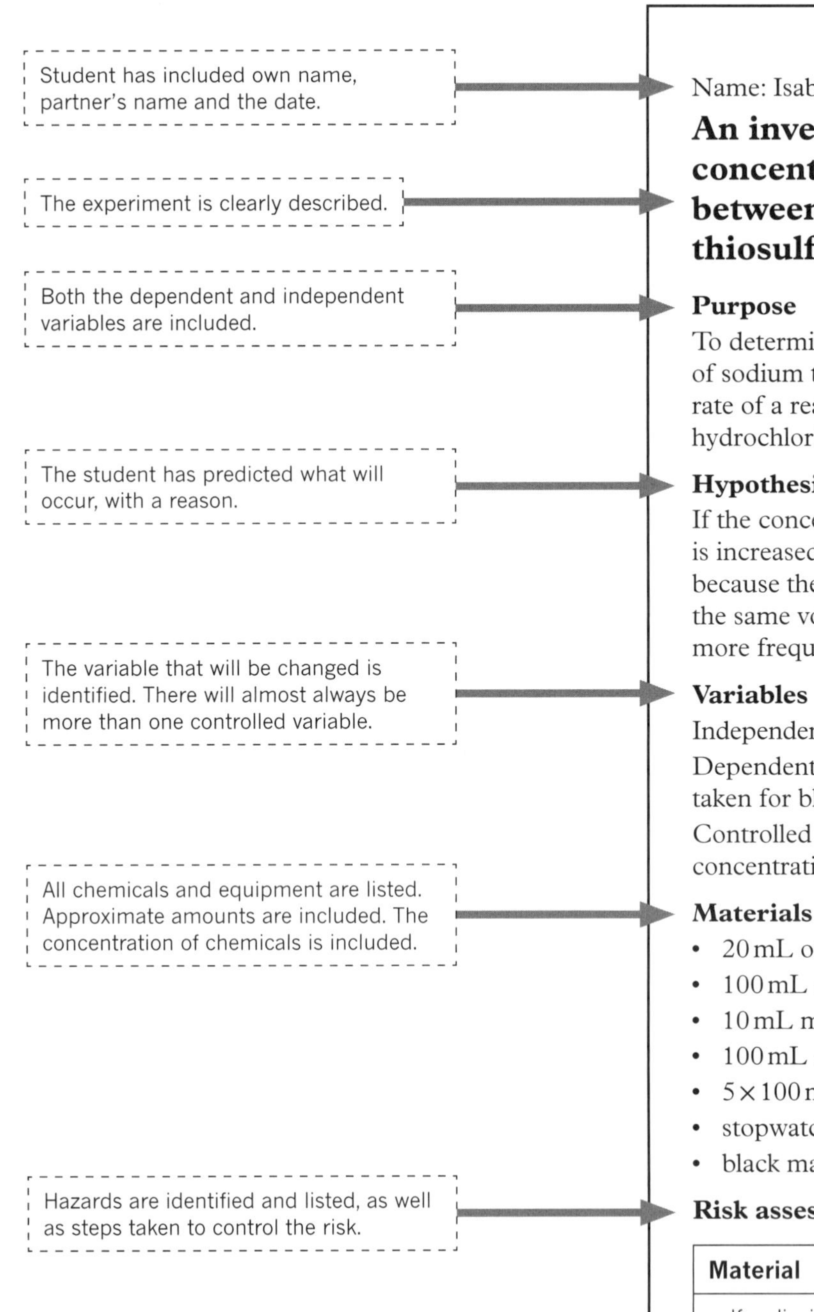

Name: Isabella Gee Partner: Li Jun Date: 23/04/17

An investigation into the effect of concentration on the rate of a reaction between hydrochloric acid and sodium thiosulfate solution

Purpose

To determine whether increasing the concentration of sodium thiosulfate solution will increase the rate of a reaction between sodium thiosulfate and hydrochloric acid.

Hypothesis

If the concentration of sodium thiosulfate solution is increased then the rate of a reaction will increase because there will be more reactant particles present in the same volume, meaning reactant particles will collide more frequently.

Variables

Independent variable: concentration of sodium thiosulfate

Dependent variable: rate of reaction (measured by time taken for black cross to disappear)

Controlled variables: volume of solutions, temperature, concentration of HCl

Materials

- 20 mL of 2.0 mol L^{-1} hydrochloric acid
- 100 mL of 0.25 mol L^{-1} sodium thiosulfate solution
- 10 mL measuring cylinder
- 100 mL measuring cylinder
- 5 × 100 mL beakers
- stopwatch
- black marking pen

Risk assessment

Material	Hazard	Control
sulfur dioxide gas	toxic by inhalation; causes burns; risk of serious damage to the eyes	Avoid breathing gas: work in a well-ventilated area and dispose of reaction mixtures using a sink in a fume cupboard. Wear safety glasses.
sulfur powder	highly flammable	Keep away from sources of ignition. Do not breathe dust. Avoid contact with eyes.
2.0 mol L^{-1} hydrochloric acid	splashes to eyes	Wear safety glasses.

 ISBN 978 1 4886 1933 5

Procedure

1 Mark a cross on a sheet of paper with a black marking pen.

2 Place a 100 mL beaker on top of the cross. Pour 10 mL of 0.25 mol L^{-1} sodium thiosulfate solution and 35 mL of water into the beaker. Add 5 mL of 2 mol L^{-1} hydrochloric acid and commence timing. Measure and record in Table 1 the time taken for the cross to disappear when viewed from above the beaker.

3 Place a 100 mL beaker on top of the cross. Pour 15 mL of 0.25 mol L^{-1} sodium thiosulfate solution and 30 mL of water into the beaker. Add 5 mL of 2 mol L^{-1} hydrochloric acid and commence timing. Measure and record the time taken for the cross to disappear when viewed from above the beaker.

4 Place another 100 mL beaker on top of the cross. Pour 25 mL of the sodium thiosulfate solution and 20 mL of water into the beaker. Add 5 mL of hydrochloric acid and commence timing. Measure and record the time taken for the cross to disappear.

5 Place a 100 mL beaker on top of the cross. Pour 35 mL of 0.25 mol L^{-1} sodium thiosulfate solution and 10 mL of water into the beaker. Add 5 mL of 2 mol L^{-1} hydrochloric acid and commence timing. Measure and record the time taken for the cross to disappear when viewed from above the beaker.

6 Place another 100 mL beaker on top of the cross. Pour 40 mL of sodium thiosulfate solution and 5 mL of water into the beaker. Add 5 mL of hydrochloric acid and commence timing. Measure and record the time taken for the cross to disappear.

7 Calculate a reaction rate according to reaction rate = $\frac{1}{\text{time}}$ taken for cross to disappear and add to Table 1.

Steps are clear and could be repeated by another student.

More than three concentrations are tested.

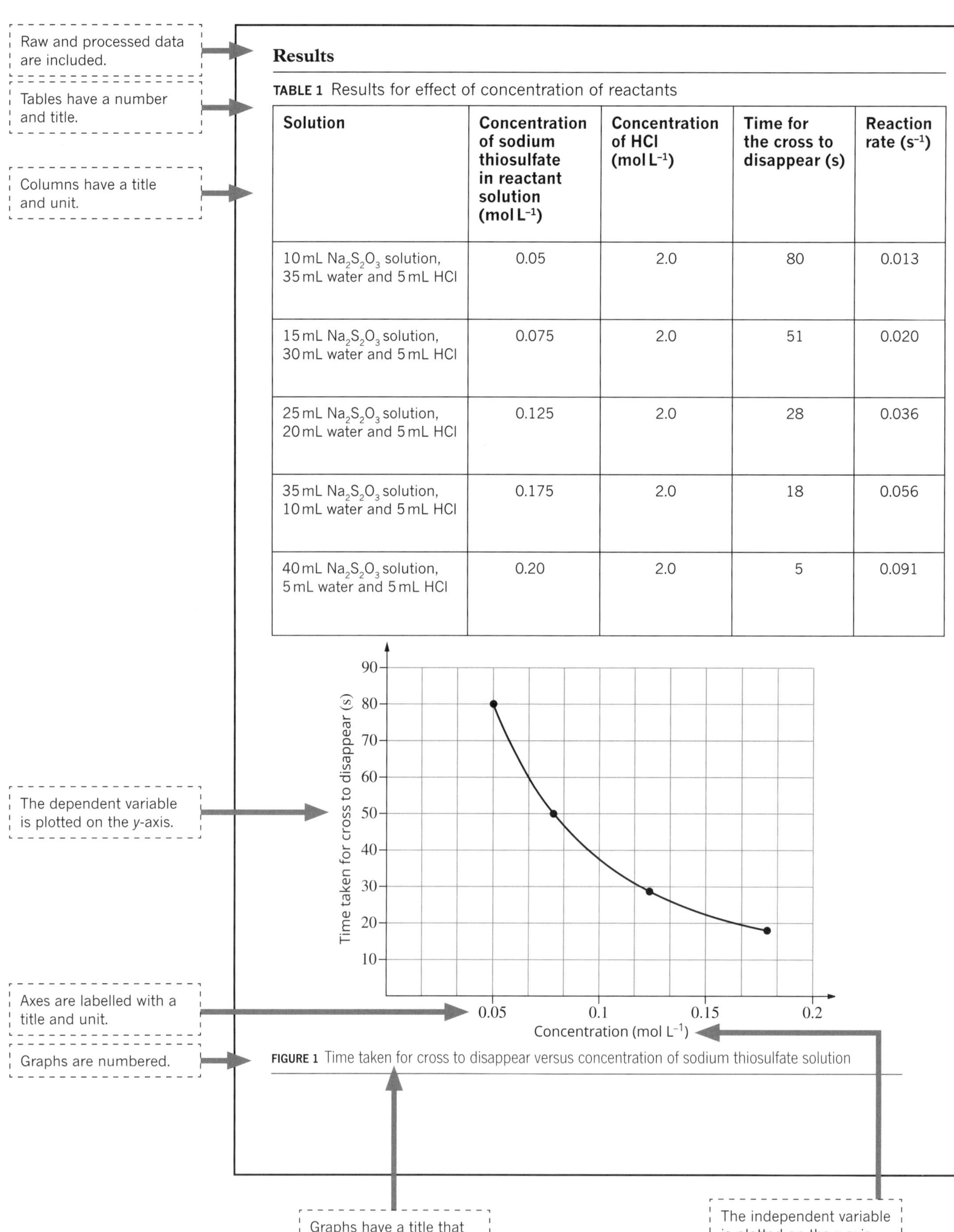

Results

TABLE 1 Results for effect of concentration of reactants

Solution	Concentration of sodium thiosulfate in reactant solution (mol L^{-1})	Concentration of HCl (mol L^{-1})	Time for the cross to disappear (s)	Reaction rate (s^{-1})
10 mL $Na_2S_2O_3$ solution, 35 mL water and 5 mL HCl	0.05	2.0	80	0.013
15 mL $Na_2S_2O_3$ solution, 30 mL water and 5 mL HCl	0.075	2.0	51	0.020
25 mL $Na_2S_2O_3$ solution, 20 mL water and 5 mL HCl	0.125	2.0	28	0.036
35 mL $Na_2S_2O_3$ solution, 10 mL water and 5 mL HCl	0.175	2.0	18	0.056
40 mL $Na_2S_2O_3$ solution, 5 mL water and 5 mL HCl	0.20	2.0	5	0.091

FIGURE 1 Time taken for cross to disappear versus concentration of sodium thiosulfate solution

ISBN 978 1 4886 1933 5

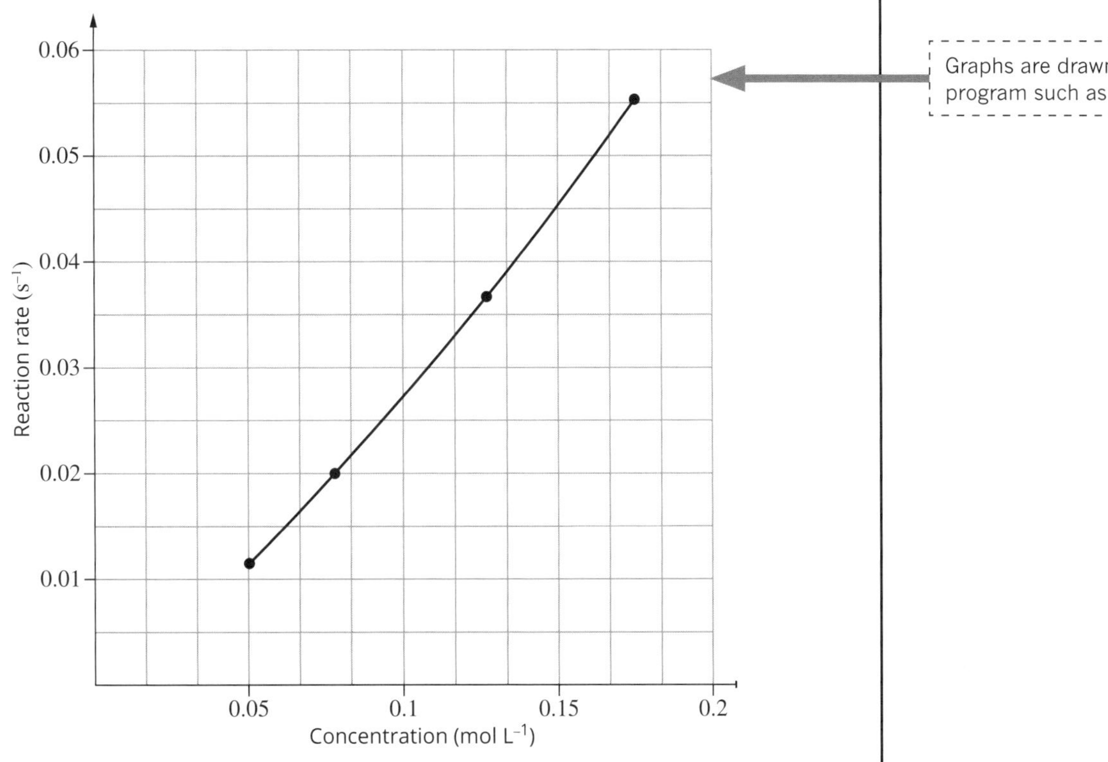

FIGURE 2 Reaction rate versus concentration of sodium thiosulfate solution

Graphs are drawn with a program such as Excel.

Discussion

The results are analysed and the experiment is evaluated.

As can be seen in Figure 2, the rate of reaction with a higher concentration of sodium thiosulfate was faster than with lower concentrations. There is evidence of a direct relationship between concentration and rate.

Results are referred to specifically.

This observation is likely due to the increased concentration of reactant particles leading to a greater frequency of collisions. This in turn means successful collisions occur more frequently and the reaction occurs at a faster rate.

An explanation is attempted.

One piece of data, for the $Na_2S_2O_3$ concentration of 0.20 mol L^{-1}, was considered to be an outlier. This error could be explained by the fact that the timer was not started until slightly after the reactants were combined. The reliability of the results would have been improved if each test was repeated three times in order to obtain an average time.

Errors are discussed.

The experiment could also be improved by increasing the range of concentration of sodium thiosulfate to see if the curve is a straight line. An alternative experiment that varied the concentration of HCl would also support whether it is the concentration of either reactants that affect the reaction rate, or whether it is the concentration of sodium thiosulfate.

Improvements are suggested.

Conclusion

Conclusion relates directly to the aim with a summarising statement.

The data showed that the reaction mixtures with a higher concentration reacted at a faster rate than those with lower concentration. For example, 0.05 mol L^{-1} $Na_2S_2O_3$ had a reaction rate of 0.013 s^{-1} and 0.175 mol L^{-1} $Na_2S_2O_3$ had a higher rate, of 0.056 s^{-1}. From this it can be concluded that increasing concentration increases the rate of a chemical reaction, thus the hypothesis has been supported.

Examples of supporting results are included.

Low standard practical report

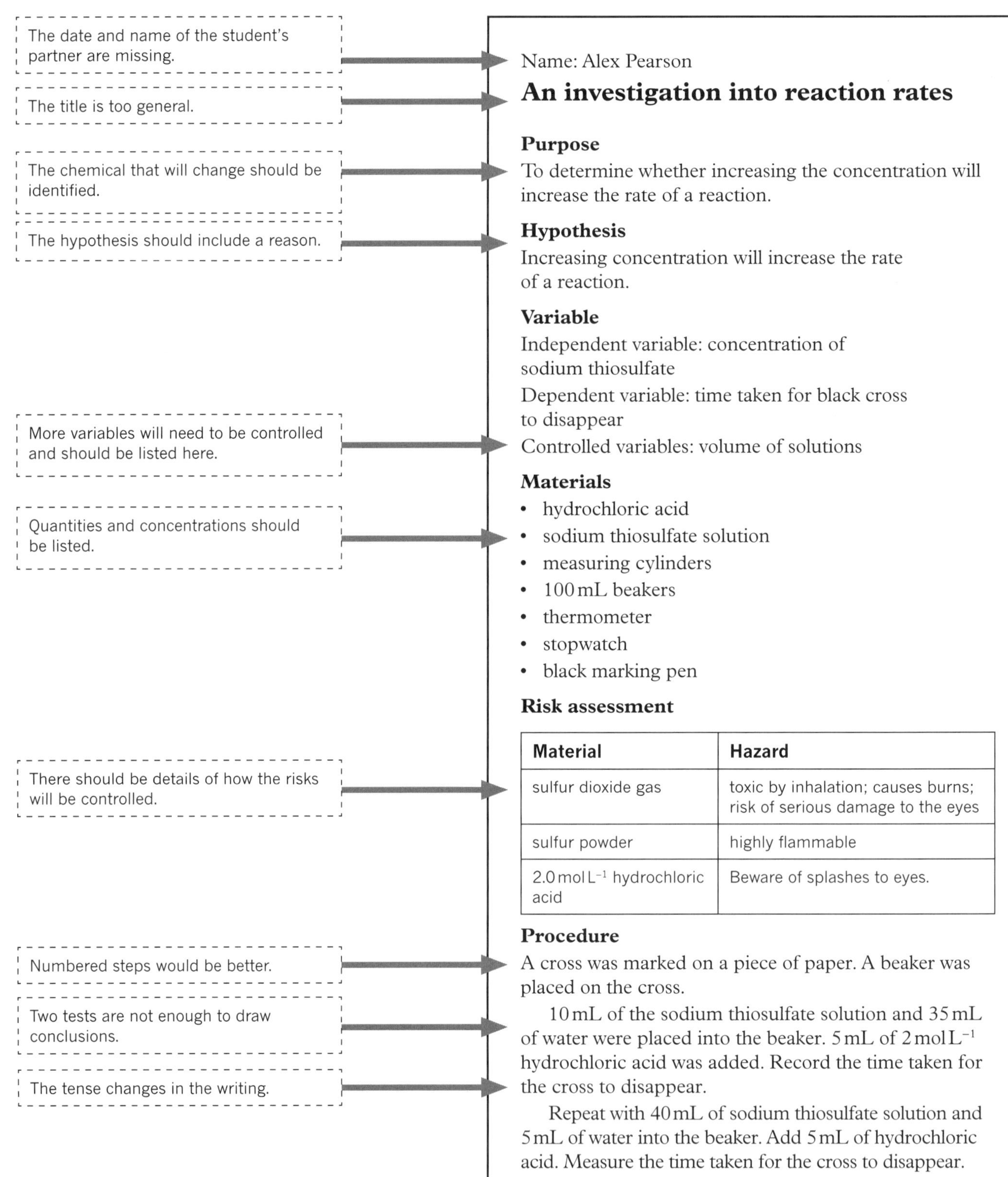

Name: Alex Pearson

An investigation into reaction rates

Purpose

To determine whether increasing the concentration will increase the rate of a reaction.

Hypothesis

Increasing concentration will increase the rate of a reaction.

Variable

Independent variable: concentration of sodium thiosulfate

Dependent variable: time taken for black cross to disappear

Controlled variables: volume of solutions

Materials

- hydrochloric acid
- sodium thiosulfate solution
- measuring cylinders
- 100 mL beakers
- thermometer
- stopwatch
- black marking pen

Risk assessment

Material	Hazard
sulfur dioxide gas	toxic by inhalation; causes burns; risk of serious damage to the eyes
sulfur powder	highly flammable
2.0 mol L^{-1} hydrochloric acid	Beware of splashes to eyes.

Procedure

A cross was marked on a piece of paper. A beaker was placed on the cross.

10 mL of the sodium thiosulfate solution and 35 mL of water were placed into the beaker. 5 mL of 2 mol L^{-1} hydrochloric acid was added. Record the time taken for the cross to disappear.

Repeat with 40 mL of sodium thiosulfate solution and 5 mL of water into the beaker. Add 5 mL of hydrochloric acid. Measure the time taken for the cross to disappear.

 ISBN 978 1 4886 1933 5

Results

Solution	Concentration of sodium thiosulfate in reactant solution (mol L^{-1})	Concentration of HCl (mol L^{-1})	Time for the cross to disappear (s)
10 mL $Na_2S_2O_3$ solution, 35 mL water and 5 mL HCl	0.05	2.0	80
40 mL $Na_2S_2O_3$ solution, 5 mL water and 5 mL HCl	0.20	2.0	5

All tables should have a heading.

Time taken for the cross to disappear is not the same as reaction rate. A further calculation would be better.

Discussion

The rate of reaction was faster for the higher concentration of sodium thiosulfate.

This observation was due to the increased concentration of reactant particles leading to a greater frequency of collisions.

Difficulties, errors and improvements should also be discussed.

Specific data should be referred to.

This explanation could be expanded.

Conclusion

The data showed that the reaction mixtures with a higher concentration reacted at a faster rate than those with lower concentration.

Results data should be referenced.

Secondary-sourced investigation

This section guides you in conducting a secondary-sourced investigation. For assistance with conducting a practical investigation, see pages ix–xvii.

A major secondary-sourced investigation is also often known as a research project. Such investigations require you to think carefully about the topic, find, collect and organise information, analyse and synthesise findings and present your ideas. The investigation process is summarised in the following flow chart. An investigation is not necessarily a straightforward linear process as shown in the flow chart. You can move back and forth between steps as needed.

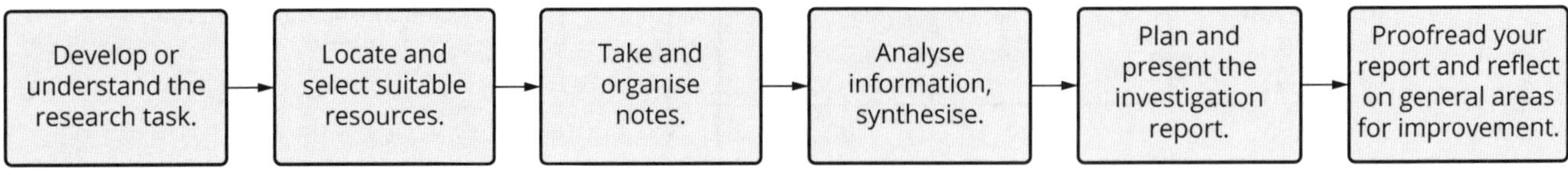

THE INVESTIGATION TASK

It is important to understand the breakdown of investigation questions or tasks. This helps you to write your own tasks and better understand the requirements of those written by others.

As you develop an investigation task, be aware of the depth of thinking it will require. The following chart provides support in writing tasks at different levels of thinking or complexity. It also provides some key words to help target tasks at these thinking levels and gives some examples of questions.

When developing your question or task, be conscious of the level at which you are pitching it. When you are writing your own questions for a depth study investigation, they should generally be at the analysis level.

Level of question complexity	Type of thinking	Words that may be used	Examples of questions
Simple	**Retrieval:** remembering, producing information on demand	• who • what • where • when • list • label • show • demonstrate • describe • name • select • state • complete • recognise • define	**1** Define the term 'molecule'. **2** What is the function of intermolecular forces? **3** List the metallic elements in order of least to most reactive. **4** What are the products of a neutralisation reaction?
	Comprehension: the ability to understand information	• who • what • where • when • explain • summarise • represent • draw • show how • describe	**1** Represent the Bohr model of the atom with a labelled diagram. **2** Why is the atomic radius of an oxygen atom smaller than the atomic radius of a lithium atom? **3** Explain why the mole concept is useful to chemists.
	Analysis: scrutinising and breaking something into its smaller parts, including: • comparing • classifying • identifying errors • concluding • predicting • judging	• why • how • categorise • compare • contrast • distinguish • sort • discriminate between • organise • deduce • generalise • critique • evaluate • diagnose • edit • identify errors • assess • identify misunderstandings • judge	**1** Categorise the following compounds according to their type of bonding. **2** Compare an exothermic and an endothermic reaction. **3** Compare and contrast alpha and beta decay.
Complex (requiring more thinking)	**Application:** using knowledge in new situations, including: • testing a hypothesis • solving a problem • experimenting and using data • decision making	• why • how • investigate • research • find out about • experiment • test • predict • solve • adapt • develop • judge • decide	**1** Many scientists believe that limiting human population growth is necessary to control environmental damage. Construct an argument for or against this statement. **2** Investigate the behaviour of a silicon diode in a simple circuit. **3** How would different water temperatures affect the rate at which a submerged iron rod rusts?

ISBN 978 1 4886 1933 5

RESOURCES

The resources you refer to in an investigation may be primary and/or secondary.

Primary sources of information are from investigations that you have conducted yourself. Examples of primary sources are results from your experiments, reports of your scientific investigations, photographs you have taken, and specimens or artefacts that you have collected. Secondary sources of information are from investigations that have been conducted by others. Secondary sources include peer-reviewed articles, textbooks, biographies, documentaries, newspaper articles and many websites.

Refer to the two tables below. The first shows features to look for when assessing and selecting the best sources for your investigation. The second shows the information that is required for the references section of your report.

Always remember to note the reference information for every resource you use. It is very time-consuming and difficult to backtrack later to obtain these details when you need them to write the references.

Selecting resources for the investigation	Tick ✔
The resource is:	
• **credible** and I can identify the author, author's expertise and publisher	
• **current** because the date of publication of the material is provided and is recent	
• **factual** and I know that it is objective material and not biased	
• **accurate** and all information is correct	
• **relevant** and covers the area I am investigating	
• **readable** and neither too simple nor too complex in its coverage of the material.	

Information required for references and examples
Article in scientific magazine Author, Initials. (year). Title of article. *Journal title, volume number*(issue number). Page numbers. Lee, M. L. (2017). Materials science: Crystals aligned through graphene. *Nature 544*(7650). 301–302.
Book Author, Initials. (year). *Title of book* (edition, if not first). City: Publisher. Rickard, G. *et al.* (2017). *Pearson Science Student Book 9.* Melbourne: Pearson Education. Where there are more than two authors, list first author then write *et al.* meaning 'and others'.
Internet Author, Initials/Name of organisation. (year). *Title of webpage or web document.* Retrieved from <URL>. Royal Society of Chemistry. (2017). *Periodic table.* Retrieved from http://www.rsc.org/periodic-table.

NOTE-TAKING AND ORGANISING NOTES

Note-taking and organising your notes takes skill. Good note-taking helps you avoid plagiarism and provides excellent information for writing up the investigation. Plagiarism is taking someone else's ideas and words and presenting them as your own work. You plagiarise if you copy sections or sentences from sources or if you cut and paste from the internet. It is acceptable to use the ideas of others but you must state clearly where the information has come from in the body of your report and include the source in your references.

The examples in the table are of original text, plagiarised text and acceptably rephrased text.

Original text	Plagiarised text	Rephrased text
Dogs have sweat glands on their feet. Dogs pant when they are hot because their sweat glands are not sufficient to cool down their bodies. In addition, their tongues allow the water from their bodies to evaporate and cool down their bodies.	Dogs have their sweat glands on their feet. When dogs get hot they pant because their sweat glands are not enough to cool down their bodies. Their tongues let the water from their bodies evaporate and cool their bodies.	Dogs pant in order to cool down. Water evaporates from their tongues and this lowers their body temperature. They can't get cool enough through just the sweat glands on their feet.

There are various approaches to effective note-taking. Whatever technique you use, try to keep notes brief, and focus on key points. Some examples include:

- dot point summary
- underlining or highlighting text
- labelled diagram
- flow chart—show sequences
- concept map—show connections between items
- Venn diagrams—show similarities and differences
- table—may incorporate any of the other note-taking techniques. Tables are useful for summarising longer and more complex information that has subparts. Adapt the table to suit your style and the task. The following sample table (partially completed) shows how this technique can be used to take notes for your secondary-sourced investigation.

Secondary-sourced investigation task Many scientists believe that limiting human population growth is necessary to control environmental damage. Construct an argument for or against this statement.			
	Source 1 (e.g. book) Title: Author: Publisher's name: Publisher's location: Date of publication:	**Source 2** (e.g. Internet) Title: Author: URL: Date accessed:	**Source 3** (e.g. science journal) Author: Date: Title of article: Journal title: Volume number: Pages:
Population growth trends		**Human population growth** Population/billions (0–8) vs Year (1750–2050)	
Impact of population growth on environment	• growth of cities • demand for resources		
Reasons to control and limit population growth	• increased demand for resources for human survival		
Reasons not to limit population growth but to allow natural population growth			• other factors contributing to environment damage; land-use policies, e.g. poor land use

SCIENTIFIC WRITING

Scientists have a particular writing style. Your investigation should use this distinctive style to communicate your ideas. Writing in the scientific style is:

- objective—describes events rather than what people think or feel
- as free as possible of bias or personal opinion
- precise—avoids exaggeration and uses qualified language
- formal—scholarly language rather than colloquial or everyday language
- concise—conveys information in short, clearly understandable sentences without unnecessary information
- simple—uses short sentences where possible
- usually written in passive voice rather than active voice, although sometimes active voice can be used
- structured to include headings, tables, diagrams, mathematical calculations.

Examples of unscientific and scientific writing are shown in the table.

Unscientific writing	Scientific writing
Subjective, biased writing: • The results were fantastic. • This produced a disgusting odour. • The breathtakingly beautiful bowerbird ...	**Objective, unbiased writing:** • The results showed ... • This produced a pungent odour ... • The golden bowerbird ...
Exaggerated writing: • The object weighed a huge amount. • The magnesium burst into huge flames. • Millions of ants swarmed over ...	**Accurate, precise writing:** • The mass of the object was 250 kg. • The magnesium burnt vigorously. • Ants swarmed all over ...

ISBN 978 1 4886 1933 5

Everyday, informal language:	**Formal language:**
• The bacteria passed away. • The results don't ... • We guessed that ... • Previous researchers were slack and missed ...	• The bacteria died. • The results do not ... • It was hypothesised that ... • Previous researchers did not perceive that ...
Active voice: • We recorded oxygen levels every hour. • We put 50 g of solute in a conical flask containing distilled water, and then we slowly added 1 mol L^{-1} hydrochloric acid.	**Passive voice:** • The oxygen level was recorded hourly. • 50 g of solute was placed in a conical flask containing distilled water, and then 1 mol L^{-1} of hydrochloric acid was slowly added.

PRESENTING THE INVESTIGATION

Scientific findings may be presented in a variety of ways. A common presentation format at science conferences is a poster. Posters can get ideas across to a large audience in an organised, concise and creative way. Other common presentation formats are essays, reports, oral presentations and articles. Each presentation format has its own conventions. The table summarises characteristics of a number of presentation formats.

Presentation formats and their characteristics

Format	Characteristics/inclusions	
Poster	• Balance of text and visuals • Title, subheadings • Balanced layout • Captions for figures and tables	• References • Hierarchy of font size according to subheading level • Consistent font style—no more than three fonts
Report/article	• Structured with introduction, paragraphs, conclusion • Includes subheadings	• Mainly text • Can include diagrams, graphs, tables
Essay	• Structured with an introduction, paragraphs, conclusion • Introduction states focus of essay • Each paragraph makes a new point supported by evidence	• Each paragraph links back to last paragraph • A text-style presentation format—visuals at end in appendix • Conclusion draws all ideas together but does not include any new information
Oral presentation	• Needs to be engaging • Use cue cards but do not read from them • Watch audience as you speak	• Stand still and don't fidget • Look at audience and appear confident

PROOFREADING

After you have completed the investigation and prepared your presentation, it is important to think about and check what you have done.

Proofread your work to minimise errors and maximise communication of the ideas from your investigation. Use these questions as a proofreading checklist.

Proofreading checklist	Tick ✔
Have I investigated the question fully?	
Have I expressed myself clearly so I communicated my ideas well?	
Have I used the scientific writing style?	
Have I included data analysis?	
Have I checked spelling, punctuation and grammar?	
Have I included references?	
Have I met the requirements of the presentation format?	

Depth study

A depth study is an investigation that allows you to look in more detail into a particular area of interest in the syllabus. A depth study gives you the chance to gain greater understanding of concepts and is designed so you take more responsibility for your own learning.

It is expected that you demonstrate the use of a variety of scientific skills. Depth studies may vary, and may include practical work, fieldwork reports, research assignments, and may be based on primary or secondary data or sources. To fulfil the minimum 15-hour time requirement, your teacher may ask you to complete one very detailed depth study or a number of smaller depth studies.

Your teacher may provide a scaffolded approach with questions that guide you through a depth study, or you may be asked to design it yourself. This section of the toolkit will help you with planning your own depth study.

PLANNING YOUR DEPTH STUDY

Careful planning of the depth study before actually starting the investigation will help you to be clear about what you will do and how you will do it. Think of yourself as taking on the role of your teacher, identifying the depth-study investigation topic, deciding how it will be investigated and presented, managing time through the investigation and checking that all syllabus requirements are met.

Clarify your ideas for your depth study, following these steps.

<table>
<tr><td>1 List all areas that interest you for your depth study. For example, transition metals, carbon nanomaterials, factors that affect the rate of a reaction, the size of a mole.</td><td>My areas of interest are:
____
____
____
____________________</td></tr>
<tr><td>2 Do some quick research into each area of interest. Identify which gives you scope to develop into a depth study.</td><td>My chosen area of investigation is:
____
____________________</td></tr>
<tr><td>3 Decide on the procedure you will use to conduct the depth study:
• a practical investigation:
- may begin with fieldwork
- may test a claim or a device
- may include data analysis
• a secondary-sourced investigation:
- may begin with fieldwork
- may be expository (explain something)
- may be an argument based on evidence
- may include data analysis
• creating:
- may be designing and constructing a working model
- may be creating a portfolio.</td><td>My procedure will be:
____
____
____
____
____
____________________</td></tr>
<tr><td>4 Decide how you will present the depth study. Select a presentation format that suits your type of investigation.
• a practical investigation (practical report)
• a secondary-sourced (research) investigation:
- documentary
- media report
- literature review
- visual presentation
- journal article
- essay
- environmental management plan
• designing or inventing or creating:
- portfolio
- working model.</td><td>I will present my depth study by:
____
____
____
____
____________________</td></tr>
</table>

ISBN 978 1 4886 1933 5

5 Now that you have thought about what and how you will conduct the depth study, think about how you can best use your time. It is tempting to spend most of the time conducting the practical or finding resources for the secondary-sourced investigation. This may leave little time to analyse data and prepare your presentation. Refer to the suggested planning timeline. Plan the depth study. — Time allocation 10% Conduct the investigation. — 40% Organise information and analyse data. — 30% Present information and proofread work. — 20%	**My depth study begins on** ____________ **and is to be completed by** ____________ Fill in the timeline to plan how you will allocate your time to complete the depth study, including specific dates.
6 Assessment requirements Depth study must: • address Working scientifically skills outcomes: Questioning and predicting, and Communicating • address a minimum of two additional Working scientifically skills outcomes: - Planning investigations - Conducting investigations - Processing data and information - Analysing data and information - Problem solving • include at least one Knowledge and understanding outcome.	**My depth study will include these Working scientifically skills:** **1** Questioning and predicting **2** Communicating **3** ____________ **4** ____________ **My depth study covers this/these knowledge and understanding outcome(s):** ____________ ____________

Study skills

There are a variety of techniques or strategies to help you study. You may find that you use different strategies in different situations. For example, you may prefer to highlight key phrases in your notebook throughout the year but make summaries of topics before an examination. The strategies you choose depend on personal preference and may not be the same as those used by classmates.

Effective study skills involve more than the learning strategies you use. Equally important is when you use skills. It is more effective to apply study skills throughout the year, revising and consolidating your knowledge as you progress through the course, rather than doing a rushed cram just before the examination. Revise work regularly. Being organised is the key to reducing your stress and setting up a study plan.

GETTING ORGANISED

To get yourself organised, try the following steps:

- Use a diary to write down all homework and assessment tasks as soon as you get them. Note due dates and what you have to do.
- Be specific about the tasks you need to do. Rather than writing 'do chemistry' it is more effective to note things such as which questions to answer and which page to look at in your student book.
- Write a list of everything you need to do each day. Tick off or cross out items as you complete them.
- Break down larger tasks into smaller, separate parts that are manageable.
- Make sure your lists and planners are realistic. Do not set yourself more than you can actually do.

STUDY TECHNIQUES

Studying requires concentration. Remove any distractions, and factor in some breaks. Allow a 10-minute break every hour. Vary your study technique depending on the content to be learned and your personal preference. Although you may have already found a study technique that works for you, also consider the options on the next page.

Study technique	Tips
Highlighting Fungi Fungi often look like plants but do not use photosynthesis. Instead they feed on dead and decaying material, breaking it down further and helping chemical elements to return to the natural environment. Mushrooms, toadstools, yeasts and moulds are different types of fungi.	• Highlight or underline key points as you read your notes or text.
Summary notes *IONIC COMPOUNDS* *Common properties of ionic compounds:* • *brittle, which means they shatter when hit with a hammer* • *hard, which means they are resistant to scratching* • *high melting point* • *unable to conduct electricity in solid state* • *good electrical conductor in liquid or aqueous states*	• Create a list of key headings and add some dot points about each heading. • Write your own summary of the key ideas in each chapter. • Use headings and subheadings. • Underline key words and key phrases. • Use simple diagrams. • The most effective chapter summaries are clear, to the point and uncluttered.
Diagrams Lowest energy orbit where electron is normally found + Higher energy orbits	• Diagrams can be used as a summary of key concepts. • They are useful memory triggers. • Diagrams cover a lot of information in a visual way, with minimal text.
Concept maps 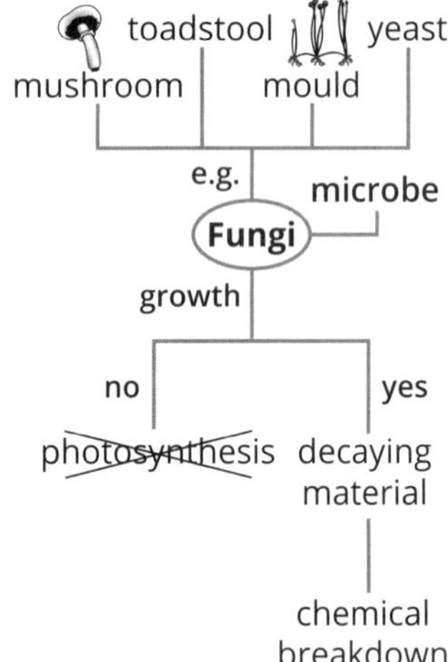	• Concept maps are a great way of connecting key terms and ideas in a simple and ordered way. • Use the lines that connect ideas to note the relationships between the words and phrases. • They may include text and images. • They may be a simple or more complex summary tool. • Concept maps and other graphic organisers such as Venn diagrams and flow charts show how information is connected and help deepen understanding.
Tables	• Tables are useful to show relationships between different factors. • Information is uncluttered.

Subatomic particles			
Particle	*Relative mass*	*Charge*	*Location*
neutron	*1*	*neutral*	*nucleus*
proton	*1*	*positive*	*nucleus*
electron	$\frac{1}{1800}$	*negative*	*orbiting nucleus*

ISBN 978 1 4886 1933 5

Mnemonic devices King Penguins Can Only Fly Going South ... triggers your memory for ... Kingdom, Phylum, Class, Order, Family, Genus, Species	• Mnemonic tools help you remember information. • To remember the classification of living things, memorise the sentence shown, which contains the initial letters for the classification types.
Glossary *Electron- negatively charged particle in an atom* *Diode- a semiconductor*	• Compile your own glossary—writing the terms and definitions will make them easier to remember. • Add images to help you remember. • Write each term in a sentence. • Memorise these terms and definitions.
Trigger words *Origin of universe- 2 theories* • *Big bang* • *Steady state theory – Fred Hoyle*	• Write down the key words associated with a topic or theme. • Trigger words are useful in helping to remember other related words and ideas.
Repeating information aloud	• This is a good way to remember and to force yourself to slow down and absorb the information.
Practice	• Do as many review questions and old exams as possible.
Flash cards *What is the Schrödinger model?* *Identify the states of matter and changes between them.*	• Making flash cards helps your understanding. • Make your own cards. • Write a question on one side and the answer on the back. • Cards can include definitions, brief explanations, diagrams, equations and graphs.
Teaching someone	• Teach friends or family members. • Teaching a difficult concept to someone means you must first understand the concept yourself.
Handwriting notes	• Hand-write rather than type summary notes. • Remember, the examination requires you to write answers. • Practise writing for long stretches of time and make sure your writing is legible.
Responding to feedback and self-correcting	• Check through all feedback from your teacher. • Highlight what was right or wrong. • Attempt to identify where you have errors and rework the answer to get it right.

EXAMINATION PREPARATION

In the weeks before the examination, begin your exam preparation. The earlier you begin revising, the easier it will be. It is also helpful to begin practising exam-style questions as early as possible, not just in the weeks ahead of the exam.

Like most skills, practice will improve your ability to do exams and to handle different types of exam question. Doing practice exams is vital because you gain experience in:

- using reading time effectively before starting to write
- allocating the right amount of time to each question
- working to a time limit
- reading and interpreting questions
- understanding what is required by each question
- planning answers
- deciding on relevant information
- proofreading/checking over your own answers
- writing efficiently for the duration of the exam.

Use the following checklist as a reminder of your study program.

Study program checklist	Tick ✔
Have I revised all areas of the course?	
Have I highlighted important points?	
Have I made a summary of the important points in each topic?	
Have I read over my revision notes?	
Have I looked at and worked through sample exam papers?	
Have I answered practice questions in the appropriate time limit?	

EXAMINATION STRATEGIES

Familiarise yourself with the conditions of the examination well before the day you sit the exam. You should know:

- the number of exams for the subject
- the amount of reading time allowed in the exam before writing begins
- the amount of writing time allocated
- any particular equipment allowed and/or required, such as a calculator, pencils, pens and ruler
- strategies to tackle the exam. Exam strategies are listed in the table.

Exam strategies
Reading time • Remember that no writing at all is allowed during this time—no note-taking, no highlighting, no underlining. • Read the instructions. • Read the short-answer questions first. • Read the multiple-choice questions next. • Read the remaining questions.
Writing time • Begin with the multiple-choice questions. • Answer every multiple-choice question, even if you can only make an educated guess. • If you are unsure of an answer to a multiple-choice question, mark it so you can come back to it if time allows. • Attempt the short-answer questions next. • Attempt the easiest short-answer questions first and work your way to the more challenging questions. • Attempt all other questions next.
Tips for answering questions • Carefully read each question, underlining key words. • Be aware that most questions are structured so they become more challenging towards the end. You may not be able to answer the last part of a question but you can earn most of the marks by answering the easier parts of the question. • Look carefully at any diagrams, pictures, tables and graphs and make sure you understand their relevance to the questions involved. • For questions with graphs, read the labels on the axes carefully, so that you can establish the relationship the graph is showing. • For questions with tables, read the headings on the columns and rows carefully, so that you can analyse the content of the table effectively. • Check for the key words in a question. Highlight them but don't colour the whole question. • Plan your answers before you write, remembering to address the exam criteria. • For questions with parts, read the whole question first. This gives you an overall picture of the question. It will also help to ensure that you do not repeat yourself in subsequent parts of the question. • Make sure you actually answer the question that is asked. • Once you have answered the question, re-read your answer and then re-read the question, to ensure that you have actually answered the question. • When writing a definition, don't use the word you are defining in your definition. • If giving values from a graph, use a ruler to line up points with the axes so you can be accurate, and always include units in your answer. • Be sure to attempt all questions. • Read over your answers to pick up careless errors—the mind is faster than the hand, and you may not always write what you intend (especially when you have limited time). • Write legibly. Exams are scanned and marked online. If the assessor can't read your answer, they can't mark it as correct. • Keep an eye on the time. • *Never* leave an exam early. Use any spare time to re-read and check your answers.
Exam cues • The number of marks allocated to a question provides a clue about how much you are expected to write. Two marks usually means you need to make a minimum of two points. • The number of lines allowed for the answer indicates the length of the expected answer. If your writing is large, you may need to turn the page and continue on the back. Make sure you indicate that the examiner must turn to the back of the page for the rest of the answer. Where the answer continues, clearly state that it is the continuation of the question and state the question number.

ISBN 978 1 4886 1933 5

MODULE 1

Properties and structure of matter

Outcomes

By the end of this module you will be able to:

- design and evaluate investigations in order to obtain primary and secondary data and information (CH11-2)
- conduct investigations to collect valid and reliable primary and secondary data and information (CH11-3)
- select and process appropriate qualitative and quantitative data and information using a range of appropriate media (CH11-4)
- communicate scientific understanding using suitable language and terminology for a specific audience or purpose (CH11-7)
- explore the properties and trends in the physical, structural and chemical aspects of matter (CH11-8)

Content

PROPERTIES OF MATTER

INQUIRY QUESTION **How do the properties of substances help us to classify and separate them?**

By the end of this module you will be able to:

- explore homogeneous mixtures and heterogeneous mixtures through practical investigations:
  - using separation techniques based on physical properties (ACSCH026)
  - calculating percentage composition by weight of component elements and/or compounds (ACSCH007)
- investigate the nomenclature of inorganic substances using International Union of Pure and Applied Chemistry (IUPAC) naming conventions
- classify the elements based on their properties and position in the periodic table through their:
  - physical properties
  - chemical properties ICT

ATOMIC STRUCTURE AND ATOMIC MASS

INQUIRY QUESTION **Why are atoms of elements different from one another?**

By the end of this module you will be able to:

- investigate the basic structure of stable and unstable isotopes by examining:
  - their position in the periodic table
  - the distribution of electrons, protons and neutrons in the atom
  - representation of the symbol, atomic number and mass number (nucleon number) ICT
- model the atom's discrete energy levels, including electronic configuration and *spdf* notation (ACSCH017, ACSCH018, ACSCH020, ACSCH022) ICT
- calculate the relative atomic mass from isotopic composition (ACSCH024) ICT N
- investigate energy levels in atoms and ions through:
  - collecting primary data from a flame test using different ionic solutions of metals (ACSCH019) ICT

- examining spectral evidence for the Bohr model and introducing the Schrödinger model

- investigate the properties of unstable isotopes using natural and human-made radioisotopes as examples, including but not limited to:
  - types of radiation
  - types of balanced nuclear reactions

PERIODICITY

INQUIRY QUESTION **Are there patterns in the properties of elements?**

By the end of this module you will be able to:

- demonstrate, explain and predict the relationships in the observable trends in the physical and chemical properties of elements in periods and groups in the periodic table, including but not limited to:
  - state of matter at room temperature
  - electronic configurations and atomic radii
  - first ionisation energy and electronegativity
  - reactivity with water

BONDING

INQUIRY QUESTION **What binds atoms together in elements and compounds?**

By the end of this module you will be able to:

- investigate the role of electronegativity in determining the ionic or covalent nature of bonds between atoms ICT N
- investigate the differences between ionic and covalent compounds through:
  - using nomenclature, valency and chemical formulae (including Lewis dot diagrams) (ACSCH029)
  - examining the spectrum of bonds between atoms with varying degrees of polarity with respect to their constituent elements' positions on the periodic table
  - modelling the shapes of molecular substances (ACSCH056, ACSCH057)
- investigate elements that possess the physical property of allotropy ICT
- investigate the different chemical structures of atoms and elements, including but not limited to:
  - ionic networks
  - covalent networks (including diamond and silicon dioxide)
  - covalent molecular structures
  - metallic structure
- explore the similarities and differences between the nature of intermolecular and intramolecular bonds and the strength of the forces associated with each, in order to explain the:
  - physical properties of elements
  - physical properties of compounds (ACSCH020, ACSCH055, ACSCH058) ICT

Key knowledge

Properties of matter

TYPES OF MATTER

Chemistry is the study of the composition, properties and reactions of matter.

All matter is composed of **atoms**. A substance composed of only one type of atom is called an **element** and a substance composed of two or more different types of atom is called a **compound**.

Elements can be **monatomic** (existing as single atoms), be composed of clusters of atoms called **molecules** or form into large networks or **giant molecules**. Compounds can be molecules or networks. These molecules and networks are the basis of all of the materials that exist in the universe and are used in everyday life.

Matter can be classified as a pure substance or a mixture. A mixture can be further classified as **homogeneous** or **heterogeneous** (Figure 1.1).

An example of a homogeneous mixture is a **solution**. The **solute** and **solvent** particles in a solution are indistinguishable from each other and the solute is distributed evenly throughout the solution. An **aqueous** solution is formed when water is the solvent or main component in a solution.

PHYSICAL PROPERTIES AND CHANGES OF STATE

The components of matter are elements, compounds and mixtures. They may be in any of three most common states of matter: solid, liquid and gas (Figure 1.2). Liquids and gases can be referred to as **fluids** because they can be poured.

FIGURE 1.1 Matter is classified according to its components and how they are arranged.

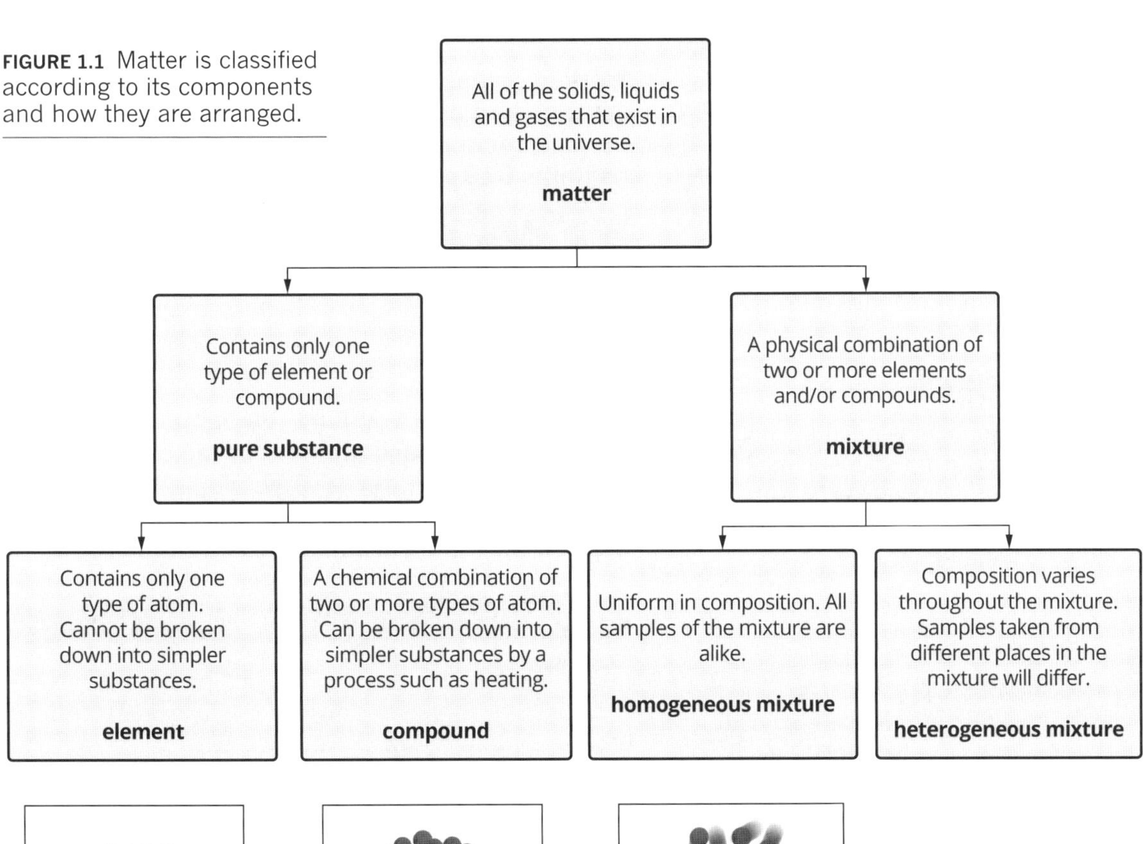

Solids are incompressible and hold their shape.

Liquids are incompressible and flow to adopt the shape of the container.

Gases are compressible and fill the shape of the container.

FIGURE 1.2 Solids, liquids and gases have different properties based on their different arrangements of particles.

Physical and chemical properties

A property is an observable characteristic of a substance. Properties can be physical or chemical:

- Physical properties can be observed or measured. Physical properties include appearance, colour, magnetism, density, brittleness or malleability, crystal shape, melting point and boiling point. Physical properties can be measured or observed without changing the substance.
- Chemical properties describe how readily a substance undergoes a chemical change. Chemical properties include the reactivity of a substance with water or acids. They also include how readily a substance will burn in oxygen and whether it is acidic or basic.

PHYSICAL AND CHEMICAL CHANGE

A **physical change** has the following characteristics:

- The physical properties of the substance are changed.
- The chemical composition of the substance is not changed.
- The change can be reversed.

The formation of a mixture from different components is a physical change because the chemical properties of each component stay the same. A physical change can also occur when a substance is heated. If it is heated enough, a change of state can occur, such as a solid melting to a liquid or a liquid evaporating to a gas. Removing the heat will reverse these physical changes.

When a **chemical change** occurs, a new substance is formed. The types of atom do not change but the atoms themselves rearrange and combine to form new substances.

SEPARATING MIXTURES

Components of a mixture can be separated by methods that use knowledge of physical properties such as magnetism, boiling point, particle size and density. The techniques most commonly used to separate mixtures in a laboratory are filtration and evaporation.

Filtration

Filtration is used to separate solids from fluids in a mixture. It uses knowledge of the physical states and the size of components in a mixture. A mixture is passed through a physical barrier called a filter. Solids are trapped by the filter while the fluid component passes through. The fluid component is called the **filtrate**. The size of the holes in the filter determines how much solid is trapped. Filtration does not completely separate the solid from the mixture. The filter can be paper or a membrane and it may have a charge. In some techniques, pressure is used to force the fluids through the filter.

Evaporation

Evaporation in the laboratory uses knowledge of the physical property of boiling point and the fact that different fluids have different boiling points. Evaporation can separate a liquid from a solid in a homogeneous solution. Boiling a solution will cause the water to evaporate, leaving behind the solute. As the solute becomes more concentrated it eventually **crystallises** out of solution. The crystals are collected by filtration or **decanting**. During decanting, a liquid is carefully poured out of a container in which solids have settled to the bottom.

Evaporation can be used to separate a liquid from another liquid using a technique called **distillation**. When the mixture is warmed, the components with the lowest boiling point will evaporate first. The vapour can be collected and separated further by more distillation.

CALCULATING PERCENTAGE COMPOSITION

Remember that compounds and mixtures are made up of more than one component:

- Compounds contain more than one element.
- Mixtures contain more than one element and/or compound.

The amount of each element in a compound or each component in a mixture can be expressed as a percentage composition by mass. It is determined using the mass of the component in a particular mass of the substance.

$$\%\text{ mass} = \frac{\text{mass of component in sample (g)}}{\text{total mass of sample (g)}} \times 100\%$$

Worked example 1.1: Find the percentage composition of copper in the copper-containing ore chalcocite. A 650 g rock of chalcocite contains 519 g of copper.

Thinking	Working
Calculate %Cu.	$\%Cu = \frac{519}{650} \times 100$ $= 79.8\%$

Worked example 1.2: Find the percentage composition of each element in the compound copper(II) carbonate. 45.0 g of copper(II) carbonate was determined to contain 23.1 g of copper, 4.37 g of carbon and 17.5 g of oxygen.

Thinking	Working
Calculate %Cu.	$\%Cu = \frac{23.1}{45.0} \times 100$ $= 51.3\%$
Calculate %C.	$\%C = \frac{4.37}{45.0} \times 100$ $= 9.71\%$
Calculate %O.	$\%O = \frac{17.5}{45.0} \times 100$ $= 38.9\%$

ELEMENTS AND THE PERIODIC TABLE

The **periodic table** is an extremely useful organisational tool for chemists. It can be used to identify patterns, trends and relationships between the structures and properties of elements. All 118 known elements are listed on the table in order of increasing atomic number, and elements that share chemical properties are grouped together. A full version of the periodic table appears later in the book. GO TO ➤ page 103

 ISBN 978 1 4886 1933 5

The modern periodic table has the following features:

- Each box of the periodic table contains one element and information about it (Figure 1.3).
- Horizontal rows are called **periods**. Periods are numbered 1–7.
- Vertical columns are called **groups**. Groups are numbered 1–18. Elements in the same group have similar chemical properties.

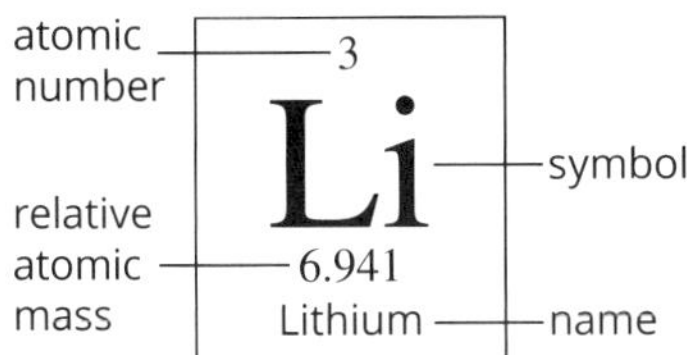

FIGURE 1.3 Periodic table information for lithium. Different periodic tables may contain different information.

The elements in the periodic table can be described as metals, non-metals or metalloids (Figure 1.4). Metallic elements and non-metallic elements are grouped because they share properties (Table 1.1). Knowing the location of an element on the periodic table can allow the prediction of some of its physical properties.

TABLE 1.1 Typical properties of metals, metalloids and non-metals

Metals	Metalloids	Non-metals
• lustrous (shiny) • malleable • ductile (can be drawn into a wire) • silvery colour • dense • high melting and boiling points • good conductors of electricity • good conductors of heat	Metalloids have some metallic and some non-metallic properties.	• dull • not malleable • not ductile • not dense • lower melting and boiling points than metals • poor conductors of electricity • poor conductors of heat

In addition to being classified as metals or non-metals, elements are placed in specific groups (columns) in the periodic table. In general, elements in the same group have similar physical and chemical properties (Table 1.2).

FIGURE 1.4 Elements are classified as metals, non-metals or metalloids in the periodic table.

KEY: Non-metals; Metals; Metalloids

atomic number — 13; symbol — Al; name — aluminium

Group

	1	2	3	4	5	6	7	8	9	10	11	12	13	14	15	16	17	18
Period 1	1 H hydrogen																	2 He helium
Period 2	3 Li lithium	4 Be beryllium											5 B boron	6 C carbon	7 N nitrogen	8 O oxygen	9 F fluorine	10 Ne neon
Period 3	11 Na sodium	12 Mg magnesium											13 Al aluminium	14 Si silicon	15 P phosphorus	16 S sulfur	17 Cl chlorine	18 Ar argon
Period 4	19 K potassium	20 Ca calcium	21 Sc scandium	22 Ti titanium	23 V vanadium	24 Cr chromium	25 Mn manganese	26 Fe iron	27 Co cobalt	28 Ni nickel	29 Cu copper	30 Zn zinc	31 Ga gallium	32 Ge germanium	33 As arsenic	34 Se selenium	35 Br bromine	36 Kr krypton
Period 5	37 Rb rubidium	38 Sr strontium	39 Y yttrium	40 Zr zirconium	41 Nb niobium	42 Mo molybdenum	43 Tc technetium	44 Ru ruthenium	45 Rh rhodium	46 Pd palladium	47 Ag silver	48 Cd cadmium	49 In indium	50 Sn tin	51 Sb antimony	52 Te tellurium	53 I iodine	54 Xe xenon
Period 6	55 Cs caesium	56 Ba barium	57–71 lanthanoids	72 Hf hafnium	73 Ta tantalum	74 W tungsten	75 Re rhenium	76 Os osmium	77 Ir iridium	78 Pt platinum	79 Au gold	80 Hg mercury	81 Tl thallium	82 Pb lead	83 Bi bismuth	84 Po polonium	85 At astatine	86 Rn radon
Period 7	87 Fr francium	88 Ra radium	89–103 actinoids	104 Rf rutherfordium	105 Db dubnium	106 Sg seaborgium	107 Bh bohrium	108 Hs hassium	109 Mt meitnerium	110 Ds darmstadtium	111 Rg roentgenium	112 Cn copernicium	113 Nh nihonium	114 Fl flerovium	115 Mc moscovium	116 Lv livermorium	117 Ts tennessine	118 Og oganesson

Lanthanides	57 La lanthanum	58 Ce cerium	59 Pr praseodymium	60 Nd neodymium	61 Pm promethium	62 Sm samarium	63 Eu europium	64 Gd gadolinium	65 Tb trebium	66 Dy dysprosium	67 Ho holmium	68 Er erbium	69 Tm thulium	70 Yb ytterbium	71 Lu lutetium
Actinides	89 Ac actinium	90 Th thorium	91 Pa protactinium	92 U uranium	93 Np neptunium	94 Pu plutonium	95 Am americium	96 Cm curium	97 Bk berkelium	98 Cf californium	99 Es einsteinium	100 Fm fremium	101 Md mendelevium	102 No nobelium	103 Lr lawrencium

TABLE 1.2 Properties of elements in some groups of the periodic table

Group number	Elements in the group	Physical and chemical properties shared by elements in that group
1	lithium (Li), sodium (Na), potassium (K), rubidium (Rb), caesium (Cs), francium (Fr)	low melting and boiling point, soft, low density, highly reactive
2	beryllium (Be), magnesium (Mg), calcium (Ca), strontium (Sr), barium (Ba), radium (Ra)	shiny, silvery-white, somewhat reactive, low density
18	helium (He), neon (Ne), argon (Ar), krypton (Kr), xenon (Xe), radon (Rn)	colourless, odourless gas at room temperature, unreactive or inert, exist as single atoms

Atomic structure and atomic mass

INSIDE ATOMS

Atoms are composed of smaller subatomic particles called **protons**, **neutrons** and **electrons**. Protons and neutrons are located in the very dense nucleus of an atom and are referred to collectively as **nucleons**. Protons, neutrons and electrons have different masses and charges (Table 1.3).

TABLE 1.3 Mass, charge and location of subatomic particles

Particle	Relative mass	Charge	Location
neutron	1	neutral	nucleus
proton	1	positive	nucleus
electron	$\frac{1}{1800}$	negative	cloud surrounding the nucleus

Atoms are electrically neutral because they contain an equal number of negative electrons and positive protons.

CLASSIFYING ATOMS

Each element is made up of one type of atom. The type of atom is determined by the number of protons in the nucleus. The **atomic number** (Z) of an element indicates the number of protons in the nucleus. The **mass number** (A) indicates the total number of nucleons (protons plus neutrons) (Figure 1.5).

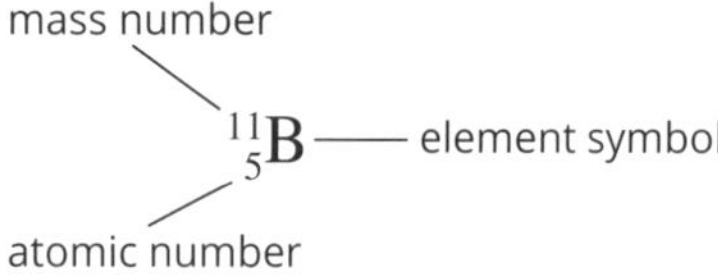

FIGURE 1.5 Representation of a boron atom

Isotopes

All atoms of the same element have identical atomic numbers. However, atoms of the same element can have different mass numbers, that is, they can have different numbers of neutrons. Atoms of the same element with different masses are called **isotopes** (Figure 1.6).

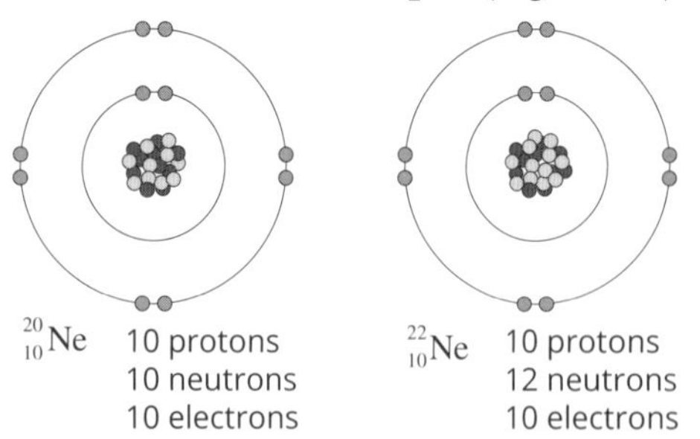

FIGURE 1.6 Isotopes of neon

Radioisotopes

Radioisotopes are isotopes that are radioactive. The nuclei of radioisotopes are unstable and will decay to form more stable nuclei. Different types of radiation can be emitted (Table 1.4).

TABLE 1.4 Types of radiation emitted during the decay of radioisotopes

Type of radiation and symbol	Energy of radiation	Description	Reason for decay
alpha (α) particle, $^{4}_{2}He$	relatively low	two protons and two neutrons; a helium nucleus	nuclei have too few neutrons to be stable
beta (β) particle, $^{0}_{-1}e$	higher than alpha particle	usually an electron, emitted from nucleus	nuclei have too many neutrons to be stable
gamma (γ) radiation	relatively high	electromagnetic radiation	unstable atom releasing energy

Balanced equations of nuclear reactions can be used to show decay of radioisotopes. **Balanced nuclear reactions** have the following features:

- The different types of radiation are represented by their symbols.
- The atomic numbers add up to the same value on both sides of the equation.
- The mass numbers add up to the same value on both sides of the equation.

For example, polonium-208 decays to produce lead-204, an alpha particle and gamma radiation. The equation is:

$$^{208}_{84}Po \rightarrow {}^{204}_{82}Pb + {}^{4}_{2}He + \gamma$$

Iodine-131 decays to produce xenon-131 and a beta particle:

$$^{131}_{53}I \rightarrow {}^{131}_{54}Xe + {}^{0}_{-1}e$$

MASSES OF PARTICLES

Masses used in chemistry are relative masses. They are relative to the standard of the common isotope **carbon-12** (^{12}C) being given a mass of exactly 12 units. Carbon-12 was selected as the standard in 1961 as a compromise between the preferences of physicists and chemists at the time. Two of the masses scientists use for different elements are:

- **relative isotopic mass** (I_r), which is the relative mass of each individual isotope of an element. Isotopes have different masses because of their different numbers of neutrons.
- **relative atomic mass** (A_r), which is the weighted average of the relative masses of the naturally occurring isotopes of a particular element. The relative atomic mass of each element is included in the periodic table.

ISBN 978 1 4886 1933 5

Determination of the relative atomic mass of an element

A **mass spectrometer** is used to identify the existence of isotopes of an element and their **relative isotopic abundances**.

The data from a mass spectrometer is displayed as a **mass spectrum** (Figure 1.7). Each peak represents one isotope. The position of each peak indicates the relative isotopic mass, and the height of the peak indicates relative isotopic abundance. The masses and relative isotopic abundances of isotopes of a particular element allow the calculation of relative atomic mass.

From the mass spectrum data in Figure 1.7:

- one isotope has a relative mass of 62.9 and relative abundance of 69.1%
- one isotope has a relative mass of 64.9 and relative abundance of 30.9%.

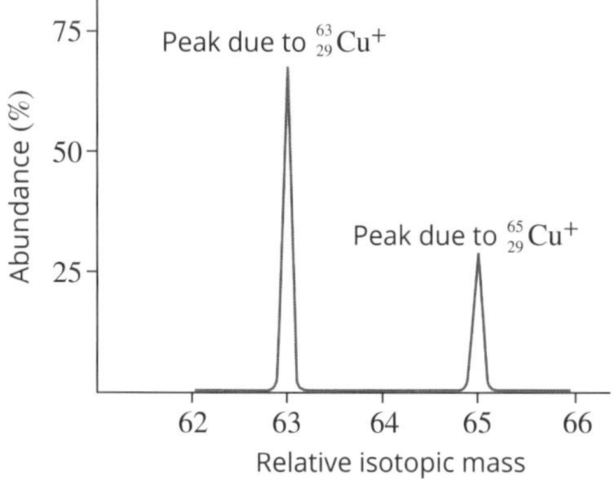

FIGURE 1.7 Mass spectrum of copper

The average relative mass of a copper atom can be calculated:

$$A_r(\text{Cu}) = \frac{(\%\text{ abundance} \times \text{relative mass}) + (\%\text{ abundance} \times \text{relative mass})}{100}$$

$$= \frac{(62.9 \times 69.1) + (64.9 \times 30.9)}{100}$$

$$= 63.5$$

ELECTRONIC STRUCTURE OF ATOMS

Emission spectra are produced when the light released by heated atoms of a particular element is passed through a prism. Each coloured line in a spectrum corresponds to light of a different energy. Emission spectra are unique to each element and are related to the electronic structure of atoms.

Neils Bohr's model of the atom explains the emission spectra of hydrogen very well. The model of the atom that existed prior to Bohr could not explain the coloured lines. In the **Bohr model**, electrons are placed into fixed orbits of specific energy (Figure 1.8):

- Electrons are able to absorb energies to move from lower to higher energy levels, causing the atom to move into an excited state.
- An excited atom is unstable so the promoted electrons immediately return to the lower energy levels. The atom returns to its ground state.
- The extra energies that the electrons had absorbed are then emitted as photons of light.
- These fixed-energy 'jumps' appear in emission spectra and are unique to each element.

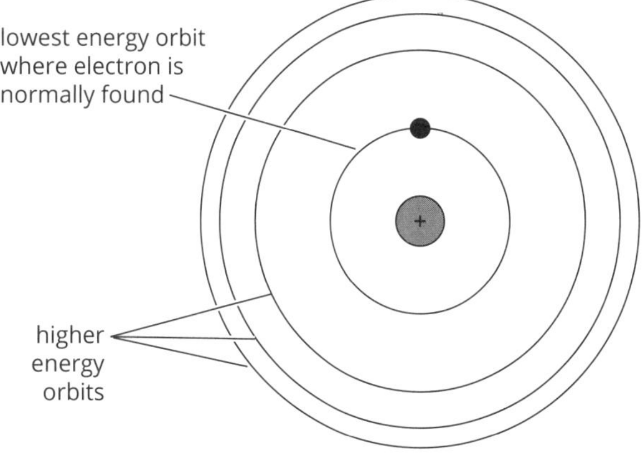

FIGURE 1.8 Bohr's model of the atom

ELECTRONIC CONFIGURATION AND THE SHELL MODEL

The energy levels of atoms are called shells. Electrons in the same shell are a similar distance from the nucleus and have similar energy. Ionisation energies were used by scientists to determine the number of electrons that can occupy each shell. Ionisation energy is the energy needed to remove an electron from an atom.

Shells are numbered 1, 2, 3, 4, 5, and so on, from the nucleus outwards. A maximum number of electrons can occupy each shell (Table 1.5).

TABLE 1.5 Maximum number of electrons that can occupy each electron shell of an atom

Electron shell number (n)	Maximum number of electrons
1	2
2	8
3	18
4	32
n	$2n^2$

Electrons fill lower energy shells before higher energy shells. An **electronic configuration** for an element lists the number of electrons in each shell.

For example, sodium has an atomic number of 11 and contains 11 electrons. Shells 1 and 2 are filled while shell 3 holds the 11th electron. The electronic configuration of sodium is 2,8,1.

The electron(s) in the outer shell of an atom are called **valence electrons**. Sodium has one valence electron.

SCHRÖDINGER MODEL OF THE ATOM

Bohr's model could not explain the emission spectra of elements with more than one electron.

Erwin Schrödinger refined Bohr's model by proposing a model of the atom using quantum mechanics. His model proposed that electrons should be regarded as having wave-like properties. According to his model, electrons are not restricted to a given orbit but behave as negative clouds of charge found in regions of space called **orbitals** (Figure 1.9).

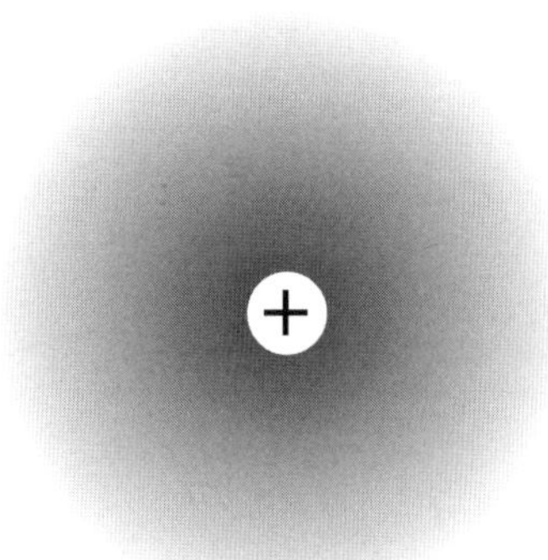

FIGURE 1.9 Electron cloud around a nucleus

The orbitals are located in subshells within shells (Table 1.6).

TABLE 1.6 Different levels of organisation of electrons

Level of organisation	Definition	Label
shell	major energy levels within an atom	1, 2, 3, 4, 5, etc.
subshell	energy levels within a shell	*s*, *p*, *d*, *f*
orbital	regions in subshells in which electrons move	

The **Pauli exclusion principle** states that each orbital may hold a maximum of two electrons:

- An *s*-subshell has 1 orbital and can hold up to 2 electrons.
- A *p*-subshell has 3 orbitals and can hold up to 6 electrons.
- A *d*-subshell has 5 orbitals and can hold up to 10 electrons.
- An *f*-subshell has 7 orbitals and can hold up to 14 electrons.

The subshells closest to the nucleus have the lowest energy and are filled first. This is known as the **Aufbau principle**.

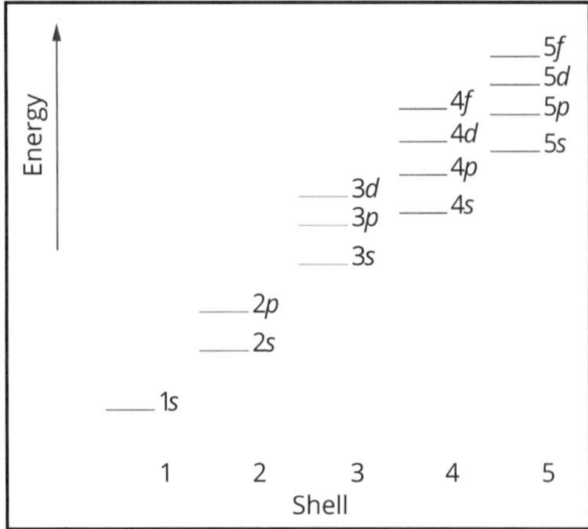

FIGURE 1.10 Energy levels in an atom

Note that some subshells are higher in energy than subshells of the next shell (Figure 1.10). For instance, the unfilled 3*d* subshell is higher in energy than the unfilled 4*s*-subshell. Therefore the 4*s* subshell is filled before the 3*d*. The electronic configurations of lithium, nitrogen, potassium and nickel are shown in Table 1.7.

TABLE 1.7 Electronic configurations of some elements

Element	Atomic number	Electronic configuration
lithium	3	$1s^2 2s^1$
nitrogen	7	$1s^2 2s^2 2p^3$
potassium	19	$1s^2 2s^2 2p^6 3s^2 3p^6 4s^1$
nickel	28	$1s^2 2s^2 2p^6 3s^2 3p^6 3d^8 4s^2$

Chromium and copper are exceptions to the usual order of filling. In these cases there is increased stability in having a half or completely full *d*-subshell. Their electronic configurations are:

Cr: $1s^2 2s^2 2p^6 3s^2 3p^6 3d^5 4s^1$

Cu: $1s^2 2s^2 2p^6 3s^2 3p^6 3d^{10} 4s^1$

Valence electrons occupy the shell furthest from the nucleus. The valence electrons determine an element's chemical properties. Atoms can lose or gain valence electrons to form charged particles called **ions**.

Periodicity

THE PERIODIC TABLE

In terms of electronic configuration, the periodic table has the following features:

- Elements in the same period of the periodic table have the same number of occupied electron shells.
- Elements in the same group have the same outer shell electronic configuration.
- Four large blocks of the periodic table can be identified (Table 1.8).

TABLE 1.8 *s*-, *p*-, *d*- and *f*-blocks in the periodic table

Block	Part of periodic table	Shell being filled
s	groups 1 and 2	*s*-subshell
p	groups 13–18	*p*-subshell
d	transition metals, groups 3–12	*d*-subshell
f	lanthanides and actinides	*f*-subshell

TRENDS IN THE PERIODIC TABLE

Trends in properties observed in the periodic table (Table 1.9) are a reflection of changing numbers of protons and electrons:

- Going down a group, the number of occupied electron shells increases by one each row so **atomic radii** are increasing and valence electrons are becoming further from the nucleus. This makes the valence electrons less tightly held so they can be more easily removed.
- Going across a period, the **core charge** of successive elements increases by one for each element (core charge = number of protons – number of inner shell electrons). As the core charge increases, the valence electrons experience a greater attraction to the nucleus and are held more tightly. They become more difficult to remove.

 ISBN 978 1 4886 1933 5

TABLE 1.9 Trends in the periodic table

Property	What it indicates	Trend going down a group	Trend going from left to right across a period
state of matter at room temperature	whether an element is a solid, liquid or gas at 25°C	There is no distinguishable trend.	Melting points increase across periods 1–3 from groups 1–14 then drastically drop for groups 15–18.
electronic configuration	number of electrons in each shell (Bohr) or subshell (Schrödinger) written in order of increasing energy	Number of occupied shells increases.	Number of valence electrons increases.
atomic radius	size of the atom	Atomic radius increases.	Atomic radius decreases.
electronegativity	ability of an atom to attract shared electrons in a molecule and how strongly the valence electrons are attracted to the nucleus	Electronegativity decreases.	Electronegativity increases.
first ionisation energy	minimum amount of energy required to remove the highest energy electron from an atom or ion and how tightly the highest energy electron is held to the nucleus	First ionisation energy decreases.	First ionisation energy increases.
reactivity with water	how readily an element will react with water, how readily a metal atom will release its outer shell electrons and how readily a non-metal atom will accept an electron	Reactivity of metals increases. Reactivity of non-metals decreases.	Reactivity of metals decreases. Reactivity of non-metals increases. However, last element in each period is an unreactive noble gas.

Bonding

Metallic elements exist as structures that contain large numbers of metal atoms bonded together. Apart from the noble gases, non-metallic elements exist as molecules in which two or more atoms are bonded together. In compounds, two or more different types of atom are bonded together. The nature of the bond is determined by the difference in the electronegativity of the two atoms involved in the bond:

- If the difference in electronegativity between the two atoms is great, an electron can be completely transferred from one atom to the other, forming an ionic bond and making an ionic compound.
- If the difference in electronegativity between the two atoms is small, the electrons are shared between the two atoms, forming a covalent bond in a molecule.

METALLIC BONDING

Remember from Table 1.1 that metals are usually:

- **lustrous** (reflective) when freshly cut or polished
- good **conductors** of heat
- good conductors of electricity
- **malleable**, which means they can be shaped by beating or rolling
- **ductile**, which means they can be drawn into a wire.

Not all metals have all of these properties. A limitation of the metallic bonding model is that it cannot explain these exceptions.

Metallic bonding model

Metal atoms generally have one, two or three electrons in their outer shell and tend to have electrons removed in their interactions with other atoms.

The **metallic bonding model** (Table 1.10) describes metals as a metallic structure which consists of a regular arrangement of stable, positively charged metal ions, called **cations**, surrounded by a 'sea' of freely moving **delocalised** valence electrons. The metal atoms achieve stability by releasing their valence electrons to become the surrounding 'sea'. **Metallic bonds** are an electrostatic attraction between the cations and the freely moving electrons.

TABLE 1.10 Summary of the metallic bonding model

Type of material	Constituent particles	Bonding model	Properties
metals (e.g. iron)	• metal cations formed when metal atoms lose their valence electrons • delocalised valence electrons	positive ions 'sea' of delocalised electrons	• generally high melting temperature • good conductor of electricity • malleable • ductile • lustrous

Structure of metallic crystals

When a metal is cooled from its liquid form, or when a metal is produced from a solution by a chemical reaction, the structure starts to form into **crystals** at many different places at once. Generally, the smaller the crystals, the less free movement of layers of ions over each other, hence the metal becomes harder and less malleable. Smaller crystals indicate more disruptions to the structure overall so the metal becomes more brittle.

IONIC BONDING

Most ionic compounds:

- are **brittle**, which means they shatter when hit with a hammer
- are **hard**, which means they are resistant to scratching
- have high melting points
- are unable to conduct electricity in solid state
- are good electrical conductors in liquid or aqueous states.

Ionic bonding model

Ionic compounds usually contain metal and non-metal atoms. The bonding model (Table 1.11) describes an ionic network that consists of a regular arrangement of metal cations and negatively charged non-metal ions called **anions**. The electronegativity difference between metal and non-metal atoms is generally high. The metal atoms fully transfer their electrons to the more electronegative non-metal atoms. **Ionic bonds** are the electrostatic attractions between the oppositely charged ions.

All models of bonding are only a representation of what is occurring at the atomic scale and so they have limitations. For example, the diagram in Table 1.11 shows the space occupied by the anions and cations but does not show the actual ionic bonds.

Ionic crystals

The arrangement of ions in a regular geometric structure is called a crystal. The exact arrangement of ions in a crystal varies according to the size and ratio of the ions.

Charges on ions

The positive or negative charge on an ion indicates whether the atom has lost or gained electrons (Table 1.12). The size of the positive or negative charge indicates the number of electrons removed or gained. The charge on an ion is called its **electrovalency**.

Ionic formulae

Anions and cations are formed when electrons are transferred from metal atoms to non-metal atoms during the formation of ionic compounds (Fig. 1.11).

TABLE 1.11 Summary of the ionic bonding model

Type of material	Constituent particles	Bonding model	Properties
ionic compounds (e.g. sodium chloride)	• cations, such as those formed when metal atoms transfer their outer shell electrons to non-metal atoms • anions, such as those formed when non-metal atoms gain electrons from metal atoms	+ cation — anion	• very high melting temperature • cannot conduct electricity when solid • can conduct electricity when molten or dissolved • brittle

TABLE 1.12 Charges on some common ions

Element	Electrons in outer shell	Tendency	Electrovalency	Ion
sodium	1	lose one electron	+1	Na^+
aluminium	3	lose three electrons	+3	Al^{3+}
oxygen	6	gain two electrons	−2	O^{2-}
chlorine	7	gain one electron	−1	Cl^-

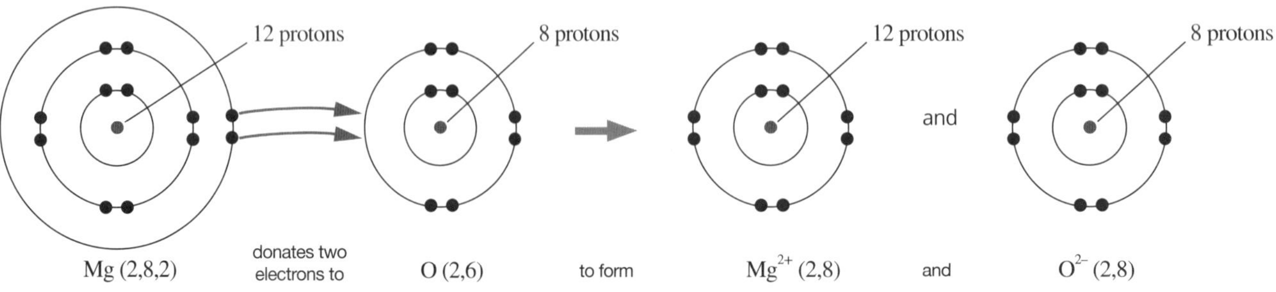

FIGURE 1.11 Transfer of electrons when magnesium reacts with oxygen

 ISBN 978 1 4886 1933 5

The ratio of metal to non-metal atoms in the compound is written in the ionic formula and depends on the electrovalencies of the ions involved (Table 1.13). The overall charge must be balanced.

TABLE 1.13 Formation of ionic compounds from ions of different electrovalencies

Cation	Anion	Formula	Name
Na^+	Cl^-	$NaCl$	sodium chloride
Mg^{2+}	Cl^-	$MgCl_2$	magnesium chloride
Ca^{2+}	NO_3^-	$Ca(NO_3)_2$	calcium nitrate
Al^{3+}	O^{2-}	Al_2O_3	aluminium oxide

COVALENT BONDING

A non-metal atom generally has five, six or seven electrons in its outer shell. When non-metal atoms bond with other non-metal atoms, their relatively similar electronegativities result in them sharing their valence electrons. **Discrete** (individual) **molecules** are formed containing fixed numbers of atoms. Molecules then interact with other molecules to give the material its observable properties.

Intramolecular forces

In general, electrons are shared between non-metal atoms so that each atom has eight electrons in its outer shell. A bond forms because the shared electrons are attracted to the positive nuclei of both atoms. Each pair of shared electrons constitutes a **covalent bond**, which is extremely strong and hard to break:

- One pair of shared electrons constitutes a single bond.
- Two pairs of shared electrons constitute a double bond.
- Three pairs of shared electrons constitute a triple bond.

The structures of molecules are most often represented by:

- **Lewis dot diagrams**, in which dots represent each valence electron (Figure 1.12a)
- structural formulae, in which a single line represents each covalent bond. Non-bonding electron pairs are not shown (Figure 1.12b).

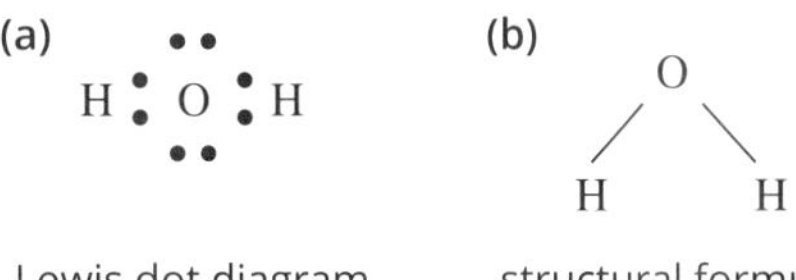

FIGURE 1.12 Different representations of a water molecule, H_2O

The four electron pairs of a stable outer shell form themselves into a **tetrahedral** shape. In this shape the pairs of negatively charged electrons have achieved maximal separation from each other within the atom according to the **valence shell electron-pair repulsion theory**. The shape of the molecule is determined by the positions of atoms. Molecules can be linear, V-shaped, pyramidal or tetrahedral (Table 1.14).

TABLE 1.14 Molecular shapes

Shape	Description	Example
linear	The two atoms of a diatomic molecule must be in a straight line.	Cl—Cl Chlorine
	The double bonds of triatomic molecules such as carbon dioxide repel each other. A straight line is as far apart as the electrons can get.	O=C=O Carbon dioxide
V-shaped (also known as angular or bent)	In a triatomic molecule, non-bonding pairs repel the bonding pairs of electrons. The bonding pairs also repel each other and try to achieve a linear shape, so a V-shape results.	O, H, H Water (showing bonding and non-bonding electron pairs) O, H, H Water (drawn without the non-bonding electron pairs, the V-shape of a water molecule is more obvious)
pyramidal	A molecule consisting of a central atom and three bonded atoms might be expected to form a flat triangular shape. These central atoms will usually have a non-bonded pair that repels the three bonded pairs, so the triangular shape becomes a pyramid.	N, H, H, H Ammonia
tetrahedral	A molecule consisting of a central atom and four bonded atoms has four sets of bonding pairs trying to get as far apart as possible. The resulting shape is tetrahedral.	H, C, H, H, H Methane

Polarity of molecules

Whether a molecule is polar or non-polar depends on the distribution of charge across the molecule. When two atoms bond that have different electronegativities, the atom with the higher electronegativity gains a greater share of the shared electron pair and thus carries a partial negative charge. The other atom will carry a partial positive charge. The molecule is called a dipole and the bond is said to be a **polar** bond (Figure 1.13).

$\overset{\delta+}{H}$——$\overset{\delta-}{Cl}$

δ+ represents a small amount of positive charge.
δ– represents a small amount of negative charge.

FIGURE 1.13 Representation of a polar bond

A polar molecule:

- contains polar bonds
- has partial charges, distributed asymmetrically across the molecule (Figure 1.14).

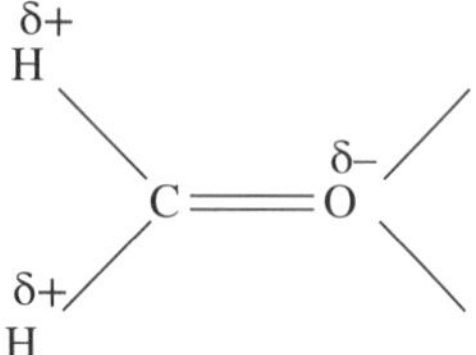

FIGURE 1.14 Structure of formaldehyde (methanal), a polar molecule

A **non-polar** molecule may (Figure 1.15) or may not (Figure 1.16) contain polar bonds. If it contains polar bonds, the partial charges are distributed symmetrically across the molecule.

FIGURE 1.15 The C–F bond in tetrafluoromethane is polar, but the molecule is non-polar.

FIGURE 1.16 Chlorine is a non-polar molecule without polar bonds.

Intermolecular forces

The common properties of covalent molecular substances include:

- low melting and boiling points due to relatively weak intermolecular forces (A covalent molecular structure can form if a covalent molecular substance is cooled enough for the molecules to form into a fixed lattice arrangement.)
- non-conduction of electricity in solid or molten states because no free-moving particles are present
- softness.

These properties can be explained by the intermolecular forces, that is, the forces between different molecules. The most significant type of intermolecular force present depends on the polarity and structure of the molecule (Table 1.15).

In contrast to the metallic or ionic bonding models, the covalent bonding model describes a number of different types of bond, each with a different strength (Figure 1.17).

TABLE 1.15 Types of intermolecular force

Force	When it is present	Bond occurs between	Relative strength
dispersion force	between all atoms/ molecules, but only relevant in non-polar molecules when no other forces are present	two different atoms or molecules	weak, but increases as the size of the atom or molecule increases
dipole–dipole bond	when molecules are polar	partially positive charge on one polar molecule and partially negative charge on another	generally stronger than dispersion forces but weaker than hydrogen bonds
hydrogen bond	when molecules are polar and have a H bonded to an N, O or F	partially positive H atom on one molecule and lone pair of electrons on N, O or F atom on another molecule	strongest intermolecular force but a lot weaker than a covalent bond

dispersion forces — dipole–dipole attractions — hydrogen bonds — covalent bonds

Increasing strength of bonds

FIGURE 1.17 Increasing strengths of different types of bonds involved in covalent bonding models

 ISBN 978 1 4886 1933 5

Covalent networks

Carbon can form networks and a range of large molecules by bonding carbon atoms in different ways. These structures are called **allotropes**. The allotropes of carbon have significantly different structures and properties.

Carbon networks

There are different types of covalent networks (Table 1.16):

- A **covalent network structure** forms when covalent bonds extend throughout an entire substance. Diamond is a carbon covalent network.
- A **covalent layer network** forms when covalent bonds extend throughout layers in a substance, and there are only weak dispersion forces between layers. Graphite is a carbon covalent layer network.

TABLE 1.16 Types of carbon network

Allotrope	Structure	Properties	Uses
diamond	covalent network structure; each carbon atom is covalently bonded to four other carbon atoms in a tetrahedral arrangement 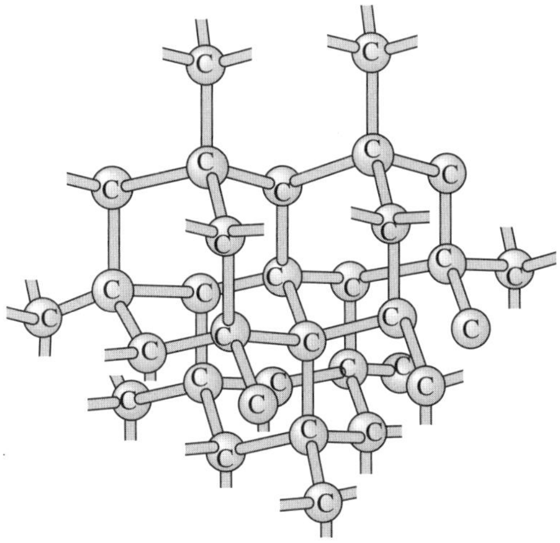	• very hard • very high melting point • non-conductor of electricity	jewellery, cutting tools, drills
graphite	covalent layer network; each carbon atom is covalently bonded to three other carbon atoms in a layer. There is one delocalised electron per carbon atom 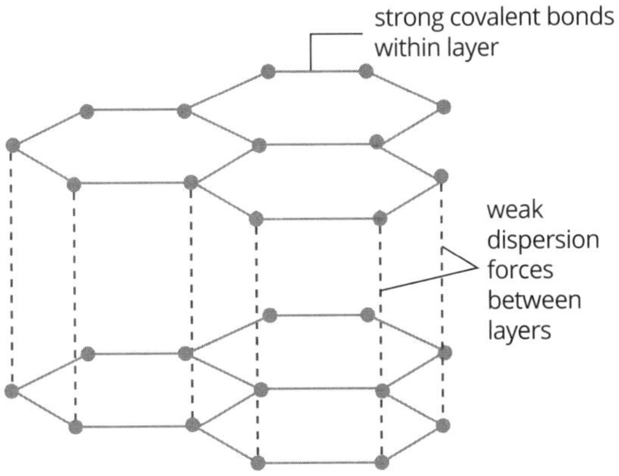 strong covalent bonds within layer weak dispersion forces between layers	• good conductor of electricity • high melting point • soft, greasy	electrodes, lubricant, pencils

ISBN 978 1 4886 1933 5

Carbon nanomaterials

Carbon nanomaterials are molecules made only of carbon atoms. These materials can be measured at the nanoscale. There are different types of carbon nanomaterials (Table 1.17).

TABLE 1.17 Types of carbon nanomaterials

Allotrope	Structure	Properties	Uses
fullerene	roughly spherical molecule in which each carbon atom is covalently bonded to three other carbon atoms. There is one delocalised electron per carbon atom	• possibility of electrical conductivity • strong • insoluble in water	• composite materials • photovoltaic cells
graphene	single layer of graphite	• good conductor of electricity • extremely strong and tough	• computer chips • circuits • electrodes • filtration of water • reinforcement

NAMING IONIC AND COVALENT COMPOUNDS

Chemists use a set of common rules to name covalent and ionic compounds, including the following:

- The least electronegative element is named first in covalent compounds; the cation is named first in ionic compounds.
- The name of the second element in covalent compounds and the anion in ionic compounds is modified by adding or altering the end of its name to 'ide'.
- Covalent compounds use the prefixes mono, di, tri, tetra, and so on, to indicate 1, 2, 3, 4, etc. to indicate the number atoms of each element.
- For ionic compounds that include metals that have ions of variable electrovalencies, the charge on the ion is specified when naming the compound by placing a Roman numeral immediately after the metal in the compound.

 ISBN 978 1 4886 1933 5

WORKSHEET 1.1

Knowledge review—thinking about matter

Examine the three atoms represented by these shell models.

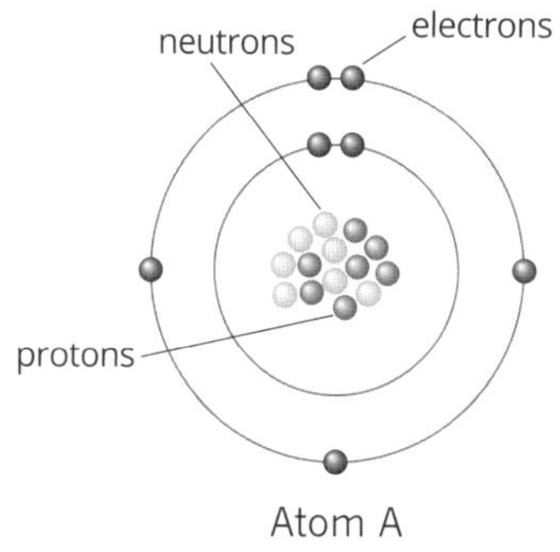

Atom A

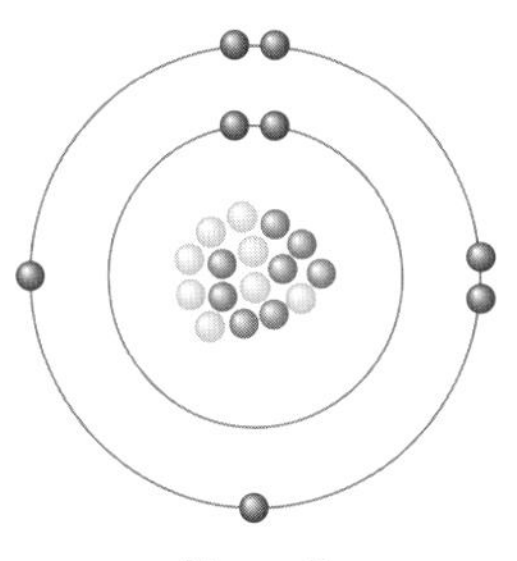
Atom B

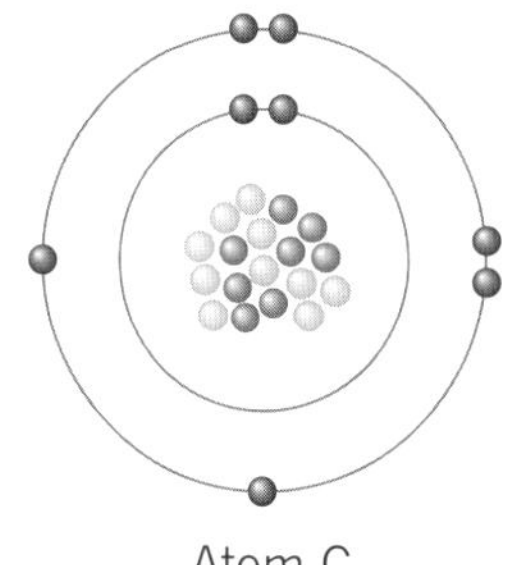
Atom C

1 Complete the information for each atom.

	Atom A	**Atom B**	**Atom C**
atomic number			
mass number			
name of element			

2 The three types of subatomic particle are labelled in atom A.

a Which subatomic particle has a negative charge? ______

b Which subatomic particle has a positive charge? ______

c Which two subatomic particles are assigned a mass of 1? ______

3 The diagrams above are representations of the structure of atoms. Describe two limitations of these diagrams.

______

______

______

4 When a chemical reaction occurs, the atoms in the reactants are rearranged to become the products. Mark each statement below about chemical reactions as true or false.

Statement about chemical reactions	**True or false?**
Mass is always conserved in a chemical reaction.	
The rate of a chemical reaction will change with changes in temperature.	
The total mass of the products may be slightly less or greater than the starting mass of the reactants.	
The number of protons in a particular atom might change during a chemical reaction.	
If there are six carbon atoms in the reactants of a chemical reaction, there must also be six carbon atoms in the products.	

WORKSHEET 1.2

Balancing protons and neutrons—radiation

1 Identify the type(s) of radiation being emitted in each of these reactions.

Nuclear reaction	Type(s) of radiation
$^{239}_{92}U \rightarrow ^{229}_{90}Th + ^{4}_{2}He$	
$^{99}_{43}Tc \rightarrow ^{99}_{43}Tc +$	
$^{238}_{92}U \rightarrow ^{234}_{90}Th + ^{4}_{2}He +$	
$^{90}_{38}Sr \rightarrow ^{90}_{39}Y + ^{0}_{-1}e$	

2 **a** Balance these nuclear reactions.

i $^{6}_{2}He \rightarrow$ ________ $+ ^{0}_{-1}e$

ii $^{146}_{62}Sm \rightarrow ^{142}_{60}Nd +$ ________

b Describe the knowledge and skills you used to balance the equation in part **a**.

3 Use the internet to find out about a human-made radioisotope that is produced for a specific purpose. Write the information you find, including a balanced nuclear reaction, in the space provided.

RATING MY LEARNING	My understanding improved	Not confident ◄──► Very confident ○ ○ ○ ○ ○	I answered questions without help	Not confident ◄──► Very confident ○ ○ ○ ○ ○	I corrected my errors without help	Not confident ◄──► Very confident ○ ○ ○ ○ ○

ISBN 978 1 4886 1933 5

WORKSHEET 1.3

Bohr to Schrödinger—spectral evidence for electronic configuration

1 Examine the emission spectra of the elements hydrogen and helium.

Emission spectrum of the hydrogen atom

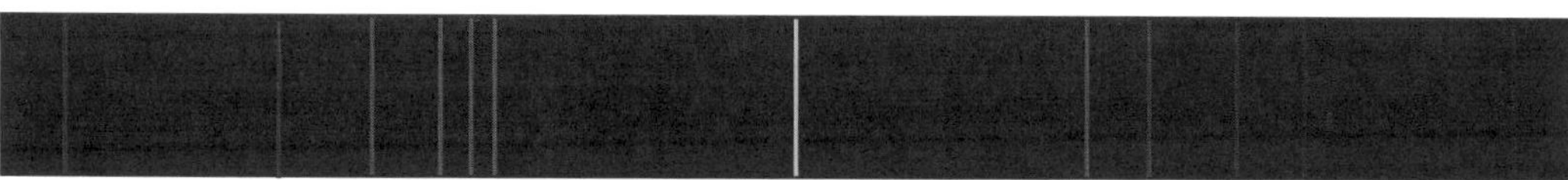

Emission spectrum of the helium atom

a Describe the Bohr model of the atom.

b Explain how the spectrum of the hydrogen atom provides evidence for the Bohr model of the atom.

c **i** What does each line represent in the helium spectrum?

ii How many different electron jumps occur in an excited helium atom as electrons return to the ground state? Explain your answer.

iii Explain why knowledge of the spectrum of helium and larger atoms led to refinement of the Bohr model and the introduction of the Schrödinger model.

ISBN 978 1 4886 1933 5

2 The order of subshell filling in the Schrödinger model is represented in the table. Fill in the number of electrons in each subshell for each element shown.

Shell	Subshell	Carbon (no. electrons = 6)	Chlorine (no. electrons = ______)	Copper (no. electrons = ______)	Vanadium (no. electrons = ______)	Selenium (no. electrons = ______)
1	*s*	2				
2	*s*	2				
	p	2				
3	*s*	0				
	p	0				
4	*s*	0				
3	*d*	0				
4	*p*	0				

3 Write the **i** electronic configuration, **ii** *spdf* notation and **iii** number of valence electrons for each element.

a carbon **i** ______ **ii** ______ **iii** ______

b chlorine **i** ______ **ii** ______ **iii** ______

c copper **i** ______ **ii** ______ **iii** ______

d vanadium **i** ______ **ii** ______ **iii** ______

e selenium **i** ______ **ii** ______ **iii** ______

4 Examine the following electronic configurations, both for the element fluorine:

species A: $1s^22s^22p^5$

species B: $1s^22s^22p^6$

Which species, A or B, represents a neutral fluorine atom? ______

Explain your answer.

______

______

5 Examine the following *spdf* notations, both for the element sodium:

species C: $1s^22s^22p^63s^1$

species D: $1s^22s^22p^53p^2$

Which species, C or D, represents an excited sodium atom? ______

Explain your answer.

______

______

RATING MY LEARNING	My understanding improved	Not confident ◄──► Very confident ○ ○ ○ ○ ○	I answered questions without help	Not confident ◄──► Very confident ○ ○ ○ ○ ○	I corrected my errors without help	Not confident ◄──► Very confident ○ ○ ○ ○ ○

 ISBN 978 1 4886 1933 5

WORKSHEET 1.4

Tracking trends—patterns in properties in the periodic table

1 Consider the elements lithium and fluorine. They are both located in period ________. Lithium is in group ________ and fluorine is in group ________.

2 Consider the following properties of elements. Circle 'increases' or 'decreases' to describe the trend from left to right across a period on the periodic table from lithium to fluorine.

Property:	Trend:		
i atomic radius	increases/decreases	**iii** first ionisation energy	increases/decreases
ii reactivity with water	increases/decreases	**iv** electronegativity	increases/decreases

3 Add electrons to the atom outlines to represent shell models of the lithium and fluorine atoms. Then write the number of protons in each nucleus and the core charge next to the atoms. Also write an electronic configuration using *spdf* notation.

Lithium atom

Fluorine atom

Number of protons: ________	Number of protons: ________
Core charge: ________	Core charge: ________
Electronic configuration: ________	Electronic configuration: ________

4 List any differences between the lithium and fluorine atoms.

5 Explain how these differences account for the different atomic radii of lithium and fluorine.

6 Explain how these differences account for the different first ionisation energies of lithium and fluorine.

7 The metallic character of an element is an indication of how many properties of a metal the elements have. The more metallic elements tend to lose their valence electrons more easily. Predict the trend for metallic character of the elements from top to bottom of a group.

RATING MY LEARNING	My understanding improved	Not confident ◄—► Very confident ○ ○ ○ ○ ○	I answered questions without help	Not confident ◄—► Very confident ○ ○ ○ ○ ○	I corrected my errors without help	Not confident ◄—► Very confident ○ ○ ○ ○ ○

WORKSHEET 1.5

The inside story on metals—metallic bonding model

1 A model of metallic bonding is shown. Describe the meaning of each label in the numbered spaces provided.

① Metallic structure:

③ Metal cation:

② Delocalised electron:

④ Electrostatic attraction:

⑤ Metallic bond:

2 Give one limitation of this representation of the metallic bonding model.

3 Next to each property of metals is a diagram. Use it to explain why metals exhibit each property.

Lustrous:

light source

reflection of light

Malleable and ductile:

Metal is hit here.

Good electrical conductor:

flow of electrons

4 Use the metallic bonding model to explain why magnesium metal has a higher melting temperature than sodium metal.

RATING MY LEARNING	My understanding improved	Not confident ◄—► Very confident ○ ○ ○ ○ ○	I answered questions without help	Not confident ◄—► Very confident ○ ○ ○ ○ ○	I corrected my errors without help	Not confident ◄—► Very confident ○ ○ ○ ○ ○

ISBN 978 1 4886 1933 5

WORKSHEET 1.6

The inside story on salts—ionic bonding model and ionic formulae

1 Natural spring water is a homogeneous solution containing a number of different dissolved ions. The table shows a typical mineral analysis from the label of a bottle of spring water.

Ion	Concentration ($mg\,L^{-1}$)
bicarbonate	5
sodium	11
chloride	20
calcium	1
magnesium	4.7
potassium	0.5

Complete this table for each ion shown.

Ion	No. electrons in outer shell of neutral atom	Charge on ion	Symbol of ion	Anion or cation?
sodium				
chloride				
calcium				
magnesium				
potassium				

2 With reference to electronic configurations, explain the following statements.

a Sodium and chlorine atoms form ions with opposite charges.

____

____

b The sodium ion has a charge of 1+ and the magnesium ion has a charge of 2+.

____

____

3 Another name for 'bicarbonate' is 'hydrogen carbonate', which is an example of a polyatomic ion.

Write the formula for the hydrogen carbonate ion. ____________

4 List the chemical formulae of all possible compounds that could be formed using the ions listed for the mineral water.

____

____

____

RATING MY LEARNING	My understanding improved	Not confident ◄—► Very confident ○ ○ ○ ○ ○	I answered questions without help	Not confident ◄—► Very confident ○ ○ ○ ○ ○	I corrected my errors without help	Not confident ◄—► Very confident ○ ○ ○ ○ ○

WORKSHEET 1.7

The inside story on molecules—covalent molecular compounds

1 Complete the following table.

Molecular formula	Elements in compound	No. electrons in outer shell of atoms of each element	No. bonds able to be formed by atoms of each element	Structural formula
HF	H	1	1	
	F	7	1	
N_2				
CO_2				
CH_3Cl				

2 Choose from the terms shown to complete the summary statements outlining the key points about electronegativity and polarity in covalent molecules. The same word may be used more than once.

distance	partial	nuclei	polar	negative	shared
polar	protons	non-polar	electronegativity	greater	

a The electrons in a covalent bond are ______________ by two atoms because they are attracted to the positive ______________ of both atoms.

b ______________ is the electron-attracting ability of an element. Different elements have different electronegativities according to the ______________ their valence electrons are from the nucleus and how many ______________ are in the nucleus. Patterns in the electronegativity of different elements are evident on the periodic table.

c In a covalent bond between atoms with different electronegativities, the atom with the ______________ electronegativity pulls the shared pair closer to its side.

d The atom with the greater share of the negatively charged electrons gains a partial ______________ charge. The other atom, which has a lesser share of the electron pair, gains a ______________ positive charge.

e The bond is said to be ______________ due to the unequal sharing of electrons.

f A ______________ molecule results when the partial charges are not spread evenly across a molecule. A ______________ molecule has either no polar bonds or has polar bonds distributed symmetrically around the molecule.

 ISBN 978 1 4886 1933 5

3 Determine the polarity and type of intermolecular force present for each molecule in the table.

Molecule showing bonding and non-bonding electron pairs	Atom with highest electronegativity	Is the bond polar? (yes or no)	Is the molecule polar? (yes or no)	Type of intermolecular force
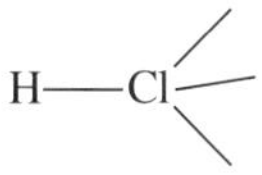 H—Cl				
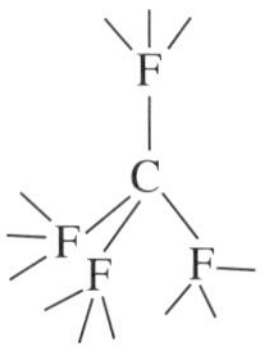 F C F F F				
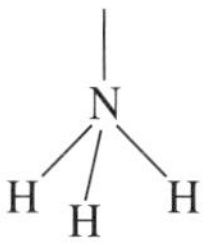 N H H H				
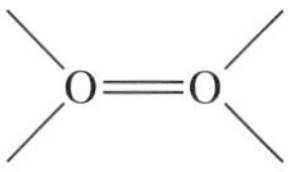 O=O				
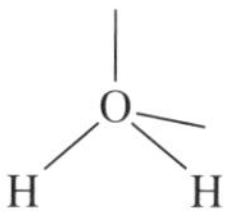 O H H				
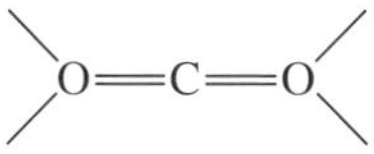 O=C=O				

RATING MY LEARNING	My understanding improved	Not confident ◄—► Very confident ○ ○ ○ ○ ○	I answered questions without help	Not confident ◄—► Very confident ○ ○ ○ ○ ○	I corrected my errors without help	Not confident ◄—► Very confident ○ ○ ○ ○ ○

WORKSHEET 1.8

Comparing structures—metallic, ionic and covalent bonding models

1 Read the definitions in the right column of the table. Write each of the following terms in the table, next to its definition.

ionic bond | cation | covalent bond | network
molecule | non-bonding electrons | metallic bond

	positively charged ion
	electrostatic attraction between a positive ion and a negative ion
	electrostatic attraction between one or more pairs of shared electrons and two positive nuclei
	regular 3D arrangement of large numbers of atoms or ions
	discrete particle containing two or more atoms covalently bonded
	valence electrons that are not involved in bonding
	electrostatic attraction between positive ions and negative delocalised electrons

2 A chemistry student is given a solid sample of an unknown material. The student carries out the following tests to determine the type of bonding present. Fill in the expected result for each test for the different models of bonding.

Test	**Metallic structure**	**Ionic network**	**Covalent molecular structure**	**Covalent network structure**
electrical conductivity of the solid samples				
electrical conductivity of the samples in molten state				
melting temperature				

RATING MY LEARNING	My understanding improved	Not confident ↔ Very confident ○ ○ ○ ○ ○	I answered questions without help	Not confident ↔ Very confident ○ ○ ○ ○ ○	I corrected my errors without help	Not confident ↔ Very confident ○ ○ ○ ○ ○

ISBN 978 1 4886 1933 5

WORKSHEET 1.9

Literacy review—naming compounds

1 Name the nitrogen-containing compounds in the table.

Molecular formula	Compound
NO	
NO_2	
NO_3	
Na_3N	
$NaNO_3$	

2 Write molecular formulae for the chlorine-containing compounds in the table.

Compound	Molecular formula
hydrogen chloride	
carbon tetrachloride	
sodium chloride	
aluminium chloride	

3 Write the rules you used to name and write the formulae of the different compounds in questions **1** and **2**.

RATING MY LEARNING	My understanding improved	Not confident ◄──► Very confident ○ ○ ○ ○ ○	I answered questions without help	Not confident ◄──► Very confident ○ ○ ○ ○ ○	I corrected my errors without help	Not confident ◄──► Very confident ○ ○ ○ ○ ○

WORKSHEET 1.10

Thinking about my learning

On completion of Module 1: Properties and structure of matter, you should be able to describe, explain and apply the relevant scientific ideas. You should be able to interpret, analyse and evaluate data.

1 The table shows the areas of key knowledge covered in the main sections of this module. Reflect on how well you understand each area. Rate your learning by shading the circle that corresponds to your level of understanding for each area. It may be helpful to use colour as a visual representation. For example:

- green—very confident
- orange—in the middle
- red—starting to develop.

Section focus	Rate my learning				
	Starting to develop ⟷ Very confident				
Properties of matter—types of matter, physical properties and changes of state, physical and chemical change	○	○	○	○	○
Analysing components of mixtures and compounds—separating mixtures and calculating percentage composition	○	○	○	○	○
Inside atoms—subatomic particles, atomic number, mass number and isotopes	○	○	○	○	○
Radio isotopes—types of radiation, balanced nuclear reactions	○	○	○	○	○
Masses of particles—relative isotopic mass, relative atomic mass, determination of relative atomic mass of an element	○	○	○	○	○
Electron structure of atoms, electronic configuration, the shell model and Schrödinger's model of the atom	○	○	○	○	○
Periodic table—trends in the physical and chemical properties of elements in periods and groups	○	○	○	○	○
Chemical structures—ionic networks, covalent lattices, covalent networks and metallic structure	○	○	○	○	○
Covalent bonding—shape and polarity of molecules and intermolecular forces	○	○	○	○	○

2 Consider points you have shaded from 'starting to develop' to 'in the middle'. List specific ideas you can identify that were challenging.

____

____

____

____

____

____

3 Write down two different strategies that you will apply to help further your understanding of these ideas.

____

____

____

____

 ISBN 978 1 4886 1933 5

PRACTICAL ACTIVITY 1.1

Separation techniques—purification of polluted water

Suggested duration: 50 minutes

INTRODUCTION

Water is never found pure in nature. It dissolves many impurities, and insoluble substances may remain in suspension. Any purification process needs to take account of the different types of substances found in water. Common water purification methods include:

- filtration: Insoluble particles are separated from water. In the laboratory, filtration can be achieved using different grades of filter paper.
- charcoal adsorption: Charcoal has the ability to adsorb many substances, that is, to hold them to its surface. Charcoal can be used to adsorb the dyes that colour water, or to adsorb substances that give water a foul odour or taste.
- distillation: Water is separated from other liquids and solids dissolved in it.
- oil–water separation: When allowed to stand undisturbed, a mixture of oil and water forms two layers, with the oil on top. The water can then be drained using a separating funnel. This method is based on the principle that oil and water have different densities and polarities, and are essentially insoluble in one another.

MATERIALS

- 100 mL polluted water
- activated charcoal
- separating funnel
- filter papers
- distillation equipment
- conductivity kit
- safety glasses

PURPOSE

- To devise and carry out a method for purification of polluted water.
- To obtain as large a volume, and as pure a sample, of water as possible from a sample of polluted water.

PRE-LAB SAFETY INFORMATION

Material used	Hazard	Control
conductivity kit	electricity	Construct the circuit with the power off. Do not touch the circuit while the power is on. Do not allow electrical equipment to come into contact with liquids.
polluted water	harmful substances in the water	Do not taste your polluted or 'purified water' sample. Wash your hands after handling polluted water.

Please indicate that you have understood the information in the safety table.

Name (print): ______________________

I understand the safety information (signature): ______________________

PROCEDURE

Record all of your procedure results in Table 1.

1. Record the volume, appearance and odour of your sample of water.
2. Using a voltage of 8 V, construct a simple circuit to test the electrical conductivity of the purified water. Be very careful not to let the electrodes touch each other.
3. Design a method of purifying your water sample. List the steps as a flow chart.

4 Identify the chemicals and equipment you will need. Show your proposed method and materials to your teacher for approval.

5 Carry out your approved procedure. Record the purpose of each step and carefully note the appearance, odour and volume of the water as you proceed.

6 When you have completed the procedure, record the appearance, odour and volume of purified water. Use the conductivity kit to determine the conductivity of the water.

7 Record the appearance, volume and conductivity of distilled water.

RESULTS

TABLE 1 Observations of water at each step of purification

Description of step	Purpose of step	Volume (mL)	Colour	Odour	Conductivity	Observations
untreated polluted water	not applicable				not applicable	
1					not applicable	
2					not applicable	
3					not applicable	
4						
distilled water						

PROCESSING DATA

Calculate the percentage yield of water at each step and record your results in Table 2.

$$\text{percentage yield} = \frac{\text{volume of water obtained (mL)}}{\text{initial volume of untreated water (mL)}} \times 100\%$$

TABLE 2 Percentage yield of water at each step

Step	Volume of water (mL)	Percentage yield
1		
2		
3		
4		

ANALYSIS OF RESULTS

1 Rate your ability to purify the water as 'very high', 'high', 'medium' or 'low'. Give a reason for your answer.

2 Which step was the most effective at purifying water? Give a reason for your answer.

 ISBN 978 1 4886 1933 5

3 What physical and/or chemical properties did you use to purify the water at each step?

4 Compare your purified water sample with those of other students. How does your sample compare in terms of:

a percentage recovery?

b apparent purity?

5 Suggest some possible reasons for the differences between your sample and those of others in your class.

6 Suggest how the purity of your sample could be further increased.

DISCUSSION

7 Find out what chemicals are added to a community's water supply in order to make it fit for drinking. How do those chemicals increase the purity of the water?

CONCLUSION

RATING MY LEARNING	My understanding improved	Not confident ◄—► Very confident ○ ○ ○ ○ ○	I answered questions without help	Not confident ◄—► Very confident ○ ○ ○ ○ ○	I corrected my errors without help	Not confident ◄—► Very confident ○ ○ ○ ○ ○

PRACTICAL ACTIVITY 1.2

Percentage composition of a compound

Suggested duration: 15 minutes

INTRODUCTION

Compounds contain two or more different types of element. The elements in each compound are present in a fixed mass ratio. In this experiment, magnesium metal is heated in oxygen to form the compound magnesium oxide. By finding the mass of the original magnesium and that of the magnesium oxide, the percentage of magnesium in magnesium oxide can be calculated.

PURPOSE

To determine the percentage by mass of magnesium in magnesium oxide.

MATERIALS

- magnesium ribbon, 5 cm long
- steel wool
- crucible and lid
- pipeclay triangle or gauze mat
- Bunsen burner
- bench mat
- tripod
- tongs
- electronic balance
- safety glasses

PRE-LAB SAFETY INFORMATION

Material used	Hazard	Control
magnesium	burns with white-hot flame that emits UV radiation; may cause eye damage	Wear eye protection; do not look directly at burning magnesium.
magnesium oxide (product)	irritant dust	Wear eye protection.

Please indicate that you have understood the information in the safety table.

Name (print): ______________________

I understand the safety information (signature): ______________________

PROCEDURE

Record all of your procedure results in Table 1.

1. Your teacher will give you a length of magnesium ribbon. Clean the magnesium thoroughly using steel wool.
2. Determine the mass of a clean, dry crucible and lid. Coil the magnesium loosely and place it in the crucible. Determine the total mass of the crucible, lid and magnesium.
3. Set up the equipment as shown in Figure 1. Heat the crucible strongly. Using tongs, occasionally lift the lid to allow air to enter the crucible but replace it quickly to avoid loss of magnesium oxide ash.
4. When the reaction appears to be complete, allow to cool to room temperature and determine the total mass of the crucible, lid and magnesium oxide.

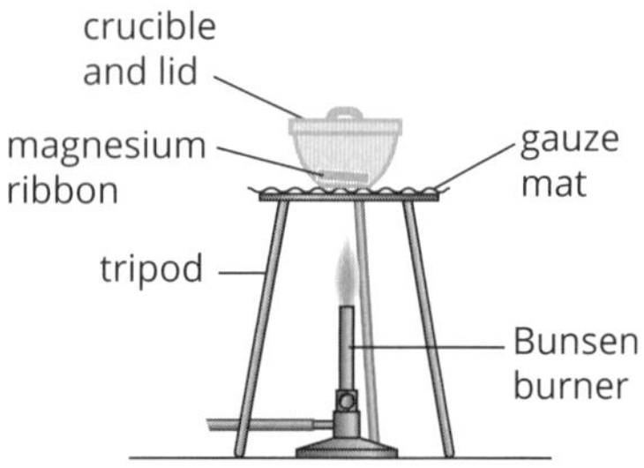

FIGURE 1 Experimental set-up

TABLE 1 Mass results

Material	Mass (g)
empty crucible and lid	
crucible, lid and magnesium	
crucible, lid and magnesium oxide	

 ISBN 978 1 4886 1933 5

PRACTICAL ACTIVITY 1.2

PROCESSING DATA

1 Calculate the mass of magnesium that reacted.

2 Calculate the mass of magnesium oxide that formed.

3 a In Table 2, record the data obtained by four other groups for their starting masses of magnesium and their masses of magnesium oxide.

TABLE 2 Group results: starting mass of magnesium and mass of magnesium oxide

Group	Mass of magnesium (g)	Mass of magnesium oxide (g)
my group		
group 1		
group 2		
group 3		
group 4		

b In Figure 2, graph the starting mass of magnesium against the mass of magnesium oxide for these five results. The starting mass of magnesium should be on the *x*-axis.

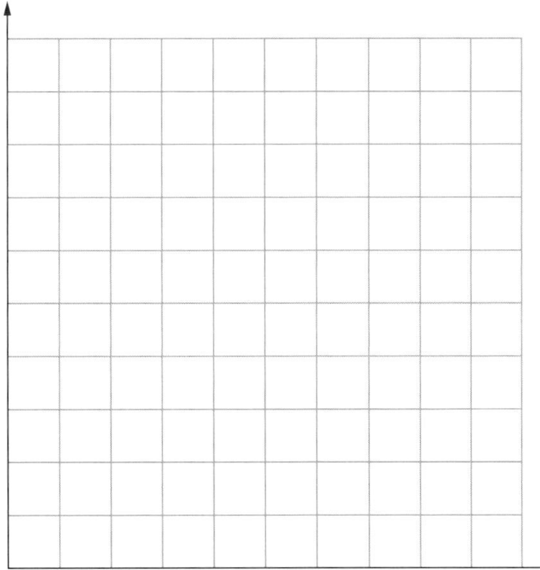

FIGURE 2 Mass of magnesium oxide produced from different starting masses of magnesium

c Is there a correlation between the starting mass of magnesium and percentage by mass of magnesium in magnesium oxide?

d Would you expect them to be similar or different? Explain your answer.

ISBN 978 1 4886 1933 5

PRACTICAL ACTIVITY 1.2

4 Calculate the percentage by mass of magnesium in magnesium oxide for your results.

5 a Calculate the percentage by mass of magnesium in magnesium oxide for the other four groups whose results you collected.

b Comment on the reliability of conclusions you can draw from this investigation.

6 Identify some of the sources of error associated with your experiment.

DISCUSSION

7 The percentage by mass of aluminium in alumina (aluminium oxide) is 52.9%. What mass of aluminium could theoretically be extracted from 800 tonnes of alumina?

CONCLUSION

RATING MY LEARNING	My understanding improved	Not confident ⟷ Very confident ○ ○ ○ ○ ○	I answered questions without help	Not confident ⟷ Very confident ○ ○ ○ ○ ○	I corrected my errors without help	Not confident ⟷ Very confident ○ ○ ○ ○ ○

 ISBN 978 1 4886 1933 5

PRACTICAL ACTIVITY 1.3

Flame colours of selected metals

Suggested duration: 20 minutes

INTRODUCTION

Distinctive colours are obtained when certain metals or their salts are heated in a flame. Heating the metals in the flame gives the electrons in their atoms enough energy to move to higher energy levels. As the electrons return to lower energy levels they give off energy as a distinctive colour.

PURPOSE

To observe some characteristic flame colours of metal ions and use the results to identify an unknown metal ion in solution.

PRE-LAB SAFETY INFORMATION

Material used	Hazard
chlorides of 0.1 mol L^{-1} sodium, potassium, calcium, strontium and copper solutions	skin irritant; may be toxic by inhalation and ingestion
0.1 mol L^{-1} barium chloride	hazardous substance; toxic if ingested

Please indicate that you have understood the information in the safety table.

Name (print): ______________________________

I understand the safety information (signature): ______________________________

MATERIALS

- 5 × 100 mL spray bottles of each of the following 0.1 mol L^{-1} solutions:
  - potassium chloride
  - calcium chloride
  - barium chloride
  - lithium chloride
  - strontium chloride
  - copper(II) chloride
  - sodium chloride
- 100 mL spray bottle containing a solution of an unknown metal chloride
- matches
- Bunsen burner and heatproof mat
- safety glasses

PROCEDURE

1. Make sure that the room is as dark as possible before starting this experiment.
2. Select a spray bottle of one of the metal solutions and, holding it about 10 cm from the Bunsen burner flame, spray into the flame.
3. Repeat this process for each of the seven solutions. Record the flame colours in Table 1.
4. Collect a spray bottle containing the unknown solution and spray the solution in the Bunsen burner flame. Record the colour in Table 1.

TABLE 1 Flame colours of different metal chlorides

Carbonate	Flame colour
potassium chloride	
calcium chloride	
barium chloride	
strontium chloride	
copper chloride	
sodium chloride	
lithium chloride	
unknown solution	

PRACTICAL ACTIVITY 1.3

DISCUSSION

1 What causes the flame colour: the cation or the anion? How can you tell?

2 Explain, in terms of the movement of electrons, why light is produced by these metals when they are heated in the Bunsen burner flame. Draw a labelled diagram as part of your answer.

3 Why does each metal produce a different flame colour?

4 Which of the metal ion solutions do you think was in the bottle labelled 'unknown'? Describe the evidence you have used.

5 Can you think of any limitations of using a flame test to determine which metal ions are present in an unknown compound?

6 Not all metals produce a coloured flame when their solutions are sprayed into the Bunsen burner flame. Considering the energy of the electrons in these metals, explain why this might be the case.

CONCLUSION

RATING MY LEARNING	My understanding improved	Not confident ◄──► Very confident ○ ○ ○ ○ ○	I answered questions without help	Not confident ◄──► Very confident ○ ○ ○ ○ ○	I corrected my errors without help	Not confident ◄──► Very confident ○ ○ ○ ○ ○

ISBN 978 1 4886 1933 5

PRACTICAL ACTIVITY 1.4

Making molecular models

Suggested duration: 50 minutes

INTRODUCTION

Molecular modelling is a useful way to represent covalent molecules in three dimensions and to relate this to the two-dimensional representations such as Lewis dot diagrams and structural formulae.

PURPOSE

To demonstrate the bonding and shape of a number of simple covalent molecules.

MATERIALS

- commercial molecular model-building kit or golf-ball-sized lumps of different colours of plasticine

PROCEDURE

1 Using the appropriately coloured 'atoms', construct a model of each molecule in tables 1–3.

2 Draw each structure with a neat diagram that shows a three-dimensional representation of your model.

3 Draw a Lewis dot diagram and structural formula for each of your molecules.

TABLE 1 Molecules with single covalent bonds

Molecule	Diagram	Lewis dot diagram	Structural formula
methane (CH_4)			
ammonia (NH_3)			
water (H_2O)			
hydrogen chloride (HCl)			
phosphine (PH_3)			
butane (C_4H_{10})			

TABLE 2 Molecules with double covalent bonds

Molecule	Diagram	Lewis dot diagram	Structural formula
oxygen (O_2)			
carbon dioxide (CO_2)			
ethene (C_2H_4)			

TABLE 3 Molecules with triple covalent bonds

Molecule	Diagram	Lewis dot diagram	Structural formula
nitrogen (N_2)			
ethyne (C_2H_2)			

PROCESSING DATA

1 Consider each molecular model. Complete Table 4 by identifying each molecule's shape, whether it is polar or non-polar and the type of intermolecular force you would expect to exist between molecules for each of these substances.

TABLE 4 Molecule shape, polarity and intermolecular force

Molecule	Shape	Polarity	Intermolecular force
methane (CH_4)			
ammonia (NH_3)			
water (H_2O)			
hydrogen chloride (HCl)			
phosphine (PH_3)			
butane (C_4H_{10})			
oxygen (O_2)			
carbon dioxide (CO_2)			
ethene (C_2H_4)			
nitrogen (N_2)			
ethyne (C_2H_2)			

ISBN 978 1 4886 1933 5

ANALYSIS OF RESULTS

2 Name the polar molecules and explain why they are polar.

3 Name at least two limitations of the models you have built in terms of how well they account for the properties of the compounds named in question **2**.

DISCUSSION

4 Name one or more elements that can form stable single, double and triple bonds with its own atoms.

5 Name two elements whose atoms never form double bonds.

6 Explain why the bonding pairs in CH_4 are arranged in a tetrahedral shape.

CONCLUSION

RATING MY LEARNING	My understanding improved	Not confident ◄——► Very confident ○ ○ ○ ○ ○	I answered questions without help	Not confident ◄——► Very confident ○ ○ ○ ○ ○	I corrected my errors without help	Not confident ◄——► Very confident ○ ○ ○ ○ ○

ISBN 978 1 4886 1933 5

PRACTICAL ACTIVITY 1.5

Comparing physical properties of three covalent networks

Suggested duration: 50 minutes

INTRODUCTION

Quartz (SiO_2), graphite (C) and dry ice, which is solid carbon dioxide (CO_2), are all composed of non-metal elements.

Silicon dioxide forms a covalent network with silicon atoms covalently bonded to oxygen atoms in a 1 : 2 ratio throughout the network. Quartz, which is essentially pure SiO_2, is colourless and not normally used as a gem. Quartz that contains various impurities produces gemstones such as amethyst (Fe impurity), rose quartz (Ti impurity) and green quartz (Au impurity).

Graphite forms a covalent layer network with carbon covalently bonded to three other carbon atoms in hexagonal rings forming layers.

Carbon dioxide consists small non-polar covalent molecules. It exists as a gas at room temperature but exists in solid form at temperatures less than −78.5°C. Solid CO_2 (dry ice) exists as a covalent molecular structure; that is, a network consisting of molecules weakly bonded to one another. The dispersion forces between carbon dioxide molecules in solid carbon dioxide are so weak that dry ice sublimes (changes from solid to gas without passing through the liquid phase).

MATERIALS

- quartz
- graphite
- dry ice
- Bunsen burner
- matches
- hammer
- 3 evaporating dishes
- conductivity kit
- sturdy bench mat or dissecting board
- tongs
- 3 × 100 mL beakers
- cold water
- universal indicator and pH chart
- insulating gloves for handling dry ice
- safety glasses

PURPOSE

- To compare the physical properties of a covalent network, a covalent layer network and a covalent molecular structure substance.
- To show that covalent molecular substances can form networks at very low temperatures, but that the intermolecular forces are very weak.

PRE-LAB SAFETY INFORMATION		
Material used	**Hazard**	**Control**
dry ice (solid carbon dioxide)	sublimes at −78.5°C; can 'burn' the skin if contact is made	Use tongs and insulating gloves when handling dry ice. Wear safety glasses and cover dry ice with a cloth when crushing it. (*Note*: Pelletised dry ice is more suitable for this experiment.)
Note: If materials are heated or burned, this should be carried out in a fume cupboard in case toxic fumes are produced.		
Please indicate that you have understood the information in the safety table. Name (print): ______ I understand the safety information (signature): ______		

PROCEDURE

1 Collect three lumps of quartz approximately 1 cm in size.

2 Collect three lengths of graphite approximately 1 cm in length. Old pencils with the outside wood removed, or pencil refills, are an excellent source of graphite.

3 Wearing safety glasses and insulating gloves, collect three lumps of dry ice approximately 2 cm in size. You may need to cover your sample of dry ice with a tea towel and hammer it to break it up into smaller pieces.

ISBN 978 1 4886 1933 5

Effect of heating

4 Place the first sample of each substance on an evaporating dish. Observe any changes over the next 15 minutes.

5 After 15 minutes, heat the sample of quartz by holding it with tongs in a Bunsen flame. Observe whether the sample melts, chars or sublimes (changes from a solid directly to the gaseous state), or whether heating has no effect.

6 Repeat step 5 with the sample of graphite.

Colour and transparency

7 Examine the second samples of quartz, graphite and dry ice. Note their colour, ability to reflect light and transparency.

Hardness, brittleness and malleability

8 Place the second samples, one at a time, on a sturdy bench mat. Lay this bench mat on the floor or on the ground outside. Put on a pair of safety glasses.

9 Tap the sample with a hammer, gently at first then with steadily increasing force. Test for hardness by observing whether the hammer marks the material. Test for brittleness by observing whether the material fractures. Test for malleability by observing whether it flattens out into a sheet.

Electrical conductivity

10 Take the third sample of each substance and use a simple conductivity kit to determine whether or not it conducts electricity.

Solubility

11 Place 60 mL water in each of three beakers. Add a few drops of universal indicator to each beaker.

12 Add a small piece (about 0.5 cm^2) of each substance to the water in each beaker. Stir and allow it to stand for 5 minutes.

13 Observe whether or not the material dissolves in water. *Note*: If the substance dissolves in water it can react further with the water, producing either a slightly acidic or basic solution. This will be indicated by any changes in colour of the universal indicator.

PROCESSING DATA

1 Complete Table 1.

TABLE 1 Physical properties of samples of quartz (SiO_2), graphite (C) and dry ice (CO_2)

Substance	Quartz (SiO_2)	Graphite (C)	Dry ice (CO_2)
diagram of structure			
name of network type			
melting properties			
colour/transparency of solid form			
electrical conductivity of solid			
solubility			

ANALYSIS OF RESULTS

2 Which solid substance is the hardest?

3 Explain why only one of the substances is able to conduct electricity.

DISCUSSION

4 Explain the different melting properties of the three types of substance.

CONCLUSION

RATING MY LEARNING	My understanding improved	Not confident ⟷ Very confident ○ ○ ○ ○ ○	I answered questions without help	Not confident ⟷ Very confident ○ ○ ○ ○ ○	I corrected my errors without help	Not confident ⟷ Very confident ○ ○ ○ ○ ○

 ISBN 978 1 4886 1933 5

DEPTH STUDY 1.1

Properties of substances—practical investigation

Suggested duration: 1 hour 45 minutes (including writing time)

INTRODUCTION

The properties of substances can indicate the nature of the particles from which the substances are made and the forces between the particles. If a substance conducts electricity, it must contain charged particles that are free to move. If a substance has a high melting temperature, there must be strong forces of attraction between the particles that make up the substance.

This depth study requires you to develop a hypothesis for a scientific investigation into the properties of different substances. You will then design and conduct an investigation, and analyse the data and information you have collected. You will communicate your ideas in a written practical report.

The questions and analysis provided here are a guide to what should be included in your report. The toolkit provides a guide to the sections and subheadings that must be included in your submitted report. GO TO ➤ page ix

PURPOSE

To determine whether the properties of a substance can be related to the elements from which it is made.

QUESTIONING AND PREDICTING

In this investigation you will use three groups of substances: those that are made only of metal elements, those that are made only of non-metal elements and those that contain both metal and non-metal elements. Use your background knowledge and the information in the introduction to develop a hypothesis for this practical investigation.

____

____

____

____

MATERIALS

- 5 cm strips of aluminium, tin and zinc
- iron nail
- approx. 8 mL of 0.1 mol L^{-1}
  - copper(II) sulfate
  - potassium iodide
  - sodium chloride
  - sucrose
- solid samples of candle wax, copper(II) sulfate, potassium iodide, sodium chloride and sucrose (approx. 5 g)
- quartz crystal
- approx. 8 mL ethanol or methylated spirits
- test-tubes and rack
- spatula
- conductivity apparatus
- 50 mL beakers
- crucibles
- gauze mat or pipeclay triangles
- Bunsen burner
- matches

PRE-LAB SAFETY INFORMATION

Material used	Hazard	Control
candle wax	flammable	Heat gently and stop heating once wax has melted. If it ignites, remove the Bunsen burner and put a lid on the crucible to extinguish the flame.
0.1 mol L^{-1} copper(II) sulfate	harmful if ingested; irritating to skin and eyes	Avoid contact. Wash hands well after use.
ethanol	flammable	Do not use near open flame.
methylated spirits	flammable; irritating to eyes, respiratory system and skin	Do not use near open flame. Avoid contact. Wash hands well after use.
potassium iodide (solid)	toxic; irritating to eyes and skin	Avoid contact. Wash hands well after use.

Please indicate that you have understood the information in the safety table.

Name (print): ______________________________

I understand the safety information (signature): ______________________________

CONDUCTING YOUR INVESTIGATION

Part A—Solubility in water

1 Design an investigation that will allow you to determine the solubility of each substance in water. Write your method in the space provided.

2 Record the materials you will require for your investigation.

3 Check your method and materials with your teacher, then carry out your investigation. Record your results in a suitable table. This will be Table 1.

 ISBN 978 1 4886 1933 5

Part B—Electrical conductivity

1 Place, in turn, samples of each of the substances in a 50 mL beaker. (Use a washed, dried beaker for each new sample.) Use enough of each substance to cover the bottom of the beaker to a depth of about 0.5 cm.

2 Test electrical conductivity of each substance for the states specified in Table 2. Substances should only be tested in the states indicated by unshaded cells in the table. Note that teacher demonstrations will be carried out for the testing of substances that have to be melted (other than candle wax). Electrical conductivity can be tested by placing the probes of a conductivity kit into the beaker or crucible containing the sample.

3 To test the conductivity of molten candle wax, place candle wax in the crucible to a depth of about 0.5 cm and heat gently over a Bunsen burner flame until the wax has melted.

4 In Table 2, record whether or not each sample tested conducts electricity.

TABLE 2 Electrical conductivity of the samples in different states

Substance	Electrical conductivity		
	In solid state	**In molten (or liquid) state**	**Dissolved in water**
aluminium			
candle wax			
copper(II) sulfate			
ethanol			
iron			
potassium iodide		teacher demonstration	
quartz crystal			
sodium chloride			
sucrose		teacher demonstration	
tin		teacher demonstration	
water			
zinc			

Part C—Melting temperature

1 Use secondary sources to find out the melting temperature of each of the substances. Enter the data into an appropriate table. Make sure you acknowledge the references you used, and use an appropriate referencing style as directed by your teacher.

PROCESSING DATA

Devise an appropriate table for inclusion in your final report that records all of the raw data you collected in parts A–C of the investigation, as well as the type(s) of element found in each substance.

ANALYSING DATA AND INFORMATION

1 Devise a table in which you can make generalisations about the solubility, electrical conductivity and melting temperature of the types of substance you tested, i.e. substances that contain only metals, substances that contain only non-metals and substances that contain both metals and non-metals. Substances with melting temperatures less than 250°C are 'low'. Substances with melting temperatures greater than 600°C are 'high'. Melting temperatures between 250°C and 600°C are 'medium'.

These questions have been designed to guide your discussion in your practical report. You should add additional analyses of your results and/or the investigation as you see fit.

2 Identify any substances that do not seem to fit with the generalisations you made in the analysis of data.

DISCUSSION

3 For a substance to conduct electricity, it must contain free-moving charged particles. Which types of substance contain free-moving charged particles?

4 Identify a type of substance that contains charged particles that are not free-moving.

5 The melting temperature of a substance depends on the strength of the forces of attraction between particles in the substance. Make generalisations about the forces of attraction between the particles in each of the three types of substance.

6 Discuss any errors or limitations in the data you have collected in this investigation.

 ISBN 978 1 4886 1933 5

7 What modifications, if any, can you make to your hypothesis based on the new evidence collected in this investigation?

8 What improvements could be made to this investigation?

COMMUNICATING

Write a practical report for this investigation in a Word document or similar. The questions you have answered will guide your report but you can also add more to your discussion. Refer to the toolkit to see what should be included in each section of a practical report. Don't forget to include a conclusion and references, especially for the secondary sources you used to research melting temperatures. **GO TO ➤** pages ix–xvii

DEPTH STUDY 1.2

Periodic variation of properties—data analysis

Suggested duration: 1 hour 45 minutes

INTRODUCTION

This depth study requires you to process and analyse data related to some properties of the first 36 elements. You will use your analysis to predict properties of bigger elements based on their position in the periodic table in relation to these elements. You will communicate your ideas in a written report.

PROCESSING DATA AND INFORMATION

Examine the data in Table 1 on the next page, which relates to properties of particular forms of the first 36 elements, then complete the tasks and answer the questions.

1 Use a spreadsheet program or graph paper to plot a graph for each of the following properties. Plot the data for each property against the atomic number of the element. Atomic number should go on the *x*-axis.

- **a** number of electrons in the outer shell
- **b** density
- **c** melting temperature
- **d** atomic radius
- **e** first ionisation energy

ANALYSING DATA AND INFORMATION

2 Examine the graphs you plotted in question **1**.

- **a** On each graph, circle the elements that belong to the same period.
- **b** Which properties of the elements display periodic variation?
- **c** Which properties do not display periodic variation?
- **d** Suggest a reason why not all properties display periodic variation.

3 Explain the trends observed in the properties that vary periodically:

- **a** from left to right across a period

ISBN 978 1 4886 1933 5

TABLE 1 Data of physical and chemical properties for the first 36 elements

Atomic number	Symbol	Electronic configuration								Relative atomic mass	Density at 298 K ($g\,mL^{-1}$)	Melting temperature (K)	Atomic radius (nm)	First ionisation energy ($MJ\,mol^{-1}$)
		1s	*2s*	*2p*	*3s*	*3p*	*3d*	*4s*	*4p*					
1	H	1								1.0	(gas)	14	0.032	1.317
2	He	2								4.0	(gas)	1	0.037	2.378
3	Li	2	1							6.9	0.53	453	0.130	0.526
4	Be	2	2							9.0	1.85	1560	0.099	0.905
5	B	2	2	1						10.8	2.34	2350	0.084	0.807
6	C	2	2	2						12.0	3.51	3773	0.075	1.092
7	N	2	2	3						14.0	(gas)	63	0.071	1.409
8	O	2	2	4						16.0	(gas)	54	0.064	1.319
9	F	2	2	5						19.0	(gas)	53	0.060	1.687
10	Ne	2	2	6						20.2	(gas)	24	0.062	2.087
11	Na	2	2	6	1					23.0	0.97	371	0.160	0.502
12	Mg	2	2	6	2					24.3	1.74	923	0.140	0.744
13	Al	2	2	6	2	1				27.0	2.70	933	0.124	0.584
14	Si	2	2	6	2	2				28.1	2.3	1687	0.114	0.793
15	P	2	2	6	2	3				31.0	1.82	317	0.109	1.017
16	S	2	2	6	2	4				32.1	2.07	388	0.104	1.006
17	Cl	2	2	6	2	5				35.5	(gas)	171	0.100	1.257
18	Ar	2	2	6	2	6				40.0	(gas)	84	0.101	1.526
19	K	2	2	6	2	6		1		39.1	0.86	336	0.200	0.425
20	Ca	2	2	6	2	6		2		40.1	1.54	1115	0.174	0.596
21	Sc	2	2	6	2	6	1	2		45.0	2.99	1814	0.159	0.637
22	Ti	2	2	6	2	6	2	2		47.9	4.5	1943	0.148	0.663
23	V	2	2	6	2	6	3	2		50.9	5.96	2183	0.144	0.657
24	Cr	2	2	6	2	6	5	1		52.0	7.20	2180	0.130	0.659
25	Mn	2	2	6	2	6	5	2		54.9	7.20	1519	0.129	0.723
26	Fe	2	2	6	2	6	6	2		55.9	7.86	1811	0.124	0.766
27	Co	2	2	6	2	6	7	2		58.9	8.9	1768	0.118	0.765
28	Ni	2	2	6	2	6	8	2		58.7	8.90	1728	0.117	0.743
29	Cu	2	2	6	2	6	10	1		63.6	8.92	1358	0.122	0.751
30	Zn	2	2	6	2	6	10	2		65.4	7.14	693	0.120	0.912
31	Ga	2	2	6	2	6	10	2	1	69.7	5.90	301	0.123	0.585
32	Ge	2	2	6	2	6	10	2	2	72.6	5.35	1211	0.120	0.766
33	As	2	2	6	2	6	10	2	3	74.9	5.73	1090	0.120	0.953
34	Se	2	2	6	2	6	10	2	4	79.0	4.81	494	0.118	0.947
35	Br	2	2	6	2	6	10	2	5	79.9	3.12	266	0.117	1.148
36	Kr	2	2	6	2	6	10	2	6	83.8	(gas)	116	0.116	1.357

b down a group.

QUESTIONING AND PREDICTING

4 a Analyse the graphs in question **1** to predict the properties of elements 37, 38 and 39.

b Use a book of chemical data, or the internet, to find the actual values for the properties of elements 37, 38 and 39. Compare the actual values with your predictions. Comment on how close your predictions are.

COMMUNICATING

The questions, data processing and analysis provided should be used as the basis for your written report of this depth study. You may also include additional areas of interest and/or research that are relevant to the topic. The following subheadings are suggested for use in your report:

- Introduction
- Processing data and information
- Analysing data and information
- Questioning and predicting
- Conclusions
- Bibliography

 ISBN 978 1 4886 1933 5

MODULE 1 • REVIEW QUESTIONS

Multiple choice

1 The isotope of iodine represented by the symbol ${}^{131}_{53}I$ is radioactive. When it decays it produces beta and gamma emissions. Which of the following best represents the particles in this isotope and the decay that occurs?

A The isotope has 53 protons, 131 neutrons and releases electrons when it decays.

B The isotope has 53 protons, 78 neutrons and releases electrons when it decays.

C The isotope has 53 neutrons, 131 protons and releases electrons when it decays.

D The isotope has 53 neutrons, 78 protons and releases helium nuclei when it decays.

2 Which of the following electronic configurations could be for a ${}^{52}_{24}Cr$ atom in an excited state?

A $1s^22s^22p^63s^23p^63d^44s^2$

B $1s^22s^22p^63s^23p^63d^4$

C $1s^22s^22p^63s^23p^63d^24s^2$

D $1s^22s^22p^63s^23p^63d^54p^1$

3 Which of the following best describes the development that the Bohr model made to atomic theory?

A Electrons orbit a central dense core called a nucleus. The atom is mostly empty space.

B Electrons exist as regions of negative cloud surrounding the nucleus.

C Electrons orbit the nucleus in fixed energy levels. The lowest energy levels are closest to the nucleus.

D Electrons are located in orbitals, which are regions of space in which electrons are located.

4 An unknown compound analysed in a laboratory contains 3.5 g of carbon, 4.8 g of oxygen and 6.5 g of other unknown substances. What is the percentage mass of oxygen in this compound?

A 0.32%

B 23%

C 32%

D 58%

5 What is the correct formula of the ionic compound made from strontium and chlorine?

A SrCl

B Sr_2Cl

C $SrCl_2$

D Sr_2Cl_2

6 What is the shape of a molecule of methane, CH_4?

A linear

B V-shaped

C pyramidal

D tetrahedral

Short answer

7 The electronegativities of C, H and Cl are 2.6, 2.1 and 3.2 respectively. What is the most polar bond out of C–Cl, C–H and H–Cl? Explain your answer.

8 Compare homogeneous and heterogeneous mixtures. Give an example of each.

Extended response

9 Briefly explain each of the following statements.

a Fluorine has a smaller atomic radius than lithium.

b Rubidium has a lower first ionisation energy than lithium.

c Metals can conduct electricity when in the solid state.

d Ionic compounds cannot conduct electricity in the solid state.

MODULE

Introduction to quantitative chemistry

Outcomes

By the end of this module you will be able to:

- design and evaluate investigations in order to obtain primary and secondary data and information (CH11-2)
- select and process appropriate qualitative and quantitative data and information using a range of appropriate media (CH11-4)
- solve scientific problems using primary and secondary data, critical thinking skills and scientific processes (CH11-6)
- describe, apply and quantitatively analyse the mole concept and stoichiometric relationships (CH11-9)

Content

CHEMICAL REACTIONS AND STOICHIOMETRY

INQUIRY QUESTION What happens in chemical reactions?

By the end of this module you will be able to:

- conduct practical investigations to observe and measure the quantitative relationships of chemical reactions, including but not limited to:
  - masses of solids and/or liquids in chemical reactions
  - volumes of gases in chemical reactions (ACSCH046) ICT N
- relate stoichiometry to the law of conservation of mass in chemical reactions by investigating:
  - balancing chemical equations (ACSCH039) N
  - solving problems regarding mass changes in chemical reactions (ACSCH046) ICT N

MOLE CONCEPT

INQUIRY QUESTION How are measurements made in chemistry?

By the end of this module you will be able to:

- conduct a practical investigation to demonstrate and calculate the molar mass (mass of one mole) of:
  - an element
  - a compound (ACSCH046) ICT N
- conduct an investigation to determine that chemicals react in simple whole number ratios by moles ICT N
- explore the concept of the mole and relate this to Avogadro's constant to describe, calculate and manipulate masses, chemical amounts and numbers of particles in: (ACSCH007, ACSCH039) ICT N
  - moles of elements and compounds $n = \frac{m}{M}$ (n = chemical amount in moles, m = mass in grams, M = molar mass in $g\,mol^{-1}$)
  - percentage composition calculations and empirical formulae
  - limiting reagent reactions

Module 2 • Introduction to quantitative chemistry

CONCENTRATION AND MOLARITY

INQUIRY QUESTION How are chemicals in solutions measured?

By the end of this module you will be able to:

- conduct practical investigations to determine the concentrations of solutions and investigate the different ways in which concentrations are measured (ACSCH046, ACSCH063) ICT N
- manipulate variables and solve problems to calculate concentration, mass or volume using:
  - $c = \frac{n}{V}$ (molarity formula) (ACSCH063) N
  - dilutions (number of moles before dilution = number of moles of sample after dilution) ICT N
- conduct an investigation to make a standard solution and perform a dilution

GAS LAWS

INQUIRY QUESTION How does the ideal gas law relate to all other gas laws?

- conduct investigations and solve problems to determine the relationship between the ideal gas law and:
  - Gay-Lussac's law (temperature)
  - Boyle's law
  - Charles' law
  - Avogadro's law (ACSCH060) ICT N

Key knowledge

Chemical reactions and stoichiometry

WRITING CHEMICAL EQUATIONS

A chemical change results in the formation of new substances. In chemical reactions, the bonds in reactants break, the particles are rearranged and products are formed. The law of conservation of mass states that matter cannot be destroyed or created in a chemical reaction. The number and types of atom that are present in the reactants must also be present in the products.

If the total mass of reactants in a chemical reaction is 10 g, then the total mass of products must also be 10 g. This does not change whether the reactants and products are solids, liquids or gases.

Consider the reaction:

zinc + hydrochloric acid → zinc chloride + hydrogen gas

The hydrogen gas bubbles out of the solution. The total mass of the reactants, zinc and hydrochloric acid, in one reaction is 26.45 g. After the bubbling due to the production of hydrogen gas stops, the mass of chemicals left in the beaker is 23.58 g. The mass of hydrogen gas released into the environment must be 26.45 − 23.58 = 2.87 g.

Balanced chemical equations

A **balanced chemical equation** represents a chemical reaction. The number of the different atoms in the reactants side is made equal to the product side. For example, hydrogen gas can react with oxygen gas to produce water. This can be represented by a word equation:

hydrogen + oxygen → water

In a chemical reaction, the names of the chemicals are replaced by their formulae:

$$H_2 + O_2 \rightarrow H_2O$$

The above equation is unbalanced. There are two oxygen atoms on the left and only one on the right of the arrow. A 2 is placed in front of the H_2O to balance the oxygen atoms. However, this means there are four hydrogen atoms on the product side so a 2 must also be placed in front of the H_2 on the reactant side. The balanced chemical equation is:

$$2H_2(g) + O_2(g) \rightarrow 2H_2O(l)$$

Note that state symbols are also in the equation to indicate the state of the reactants and products:

- (g) is for gas
- (l) is for liquid
- (s) is for solid
- (aq) is for aqueous, which means 'dissolved in water'.

The numbers in front of the chemical formulae in the reaction are called coefficients.

PROBLEMS INVOLVING CONSERVATION OF MASS

Stoichiometry allows calculation of the amounts of reactants consumed and products formed from a given amount of reactant or product. The coefficients in a balanced chemical equation provide a ratio by which atoms or molecules react and are produced.

Consider the following equation:

$$MgCl_2(aq) + 2AgNO_3(aq) \rightarrow 2AgCl(s) + Mg(NO_3)_2(aq)$$

According to the coefficients in this balanced chemical equation, the stoichiometric ratio is:

$$MgCl_2 : AgNO_3 : AgCl : Mg(NO_3)_2$$
$$1 : 2 : 2 : 1$$

That is, one $MgCl_2$ requires two $AgNO_3$ to react and they produce two AgCl and one $Mg(NO_3)_2$.

Mole concept

INTRODUCING THE MOLE

Atoms are so small that they cannot readily be counted individually or even in the thousands or millions. Instead, chemists use a special unit of measurement called the mole (Table 2.1). One mole of any substance contains **Avogadro's constant** of particles, i.e. 6.022×10^{23} particles.

It is very difficult to conceptualise just how big Avogadro's constant is, especially when atoms and molecules are so small (Figure 2.1).

TABLE 2.1 Examples of the use of the mole

Number of moles	Number and type of particles
1.00 mol of Na atoms	6.022×10^{23} Na atoms
1.00 mol of H_2O molecules	6.022×10^{23} H_2O molecules 6.022×10^{23} O atoms because there is one O atom per molecule $2 \times 6.022 \times 10^{23} = 1.204 \times 10^{24}$ H atoms because there are two H atoms per water molecule
3.00 mol of O_2 molecules	$3 \times 6.022 \times 10^{23} = 1.807 \times 10^{24}$ O_2 molecules $2 \times 1.807 \times 10^{24} = 3.613 \times 10^{24}$ O atoms

 ISBN 978 1 4886 1933 5

FIGURE 2.1 One mole of soft-drink cans would cover the Earth's surface to a depth of more than 320 km.

The mole is given the symbol n and the unit mol. A useful formula links the amount of substance (n), the number of particles in the substance (N) and Avogadro's constant (N_A) of 6.022×10^{23}.

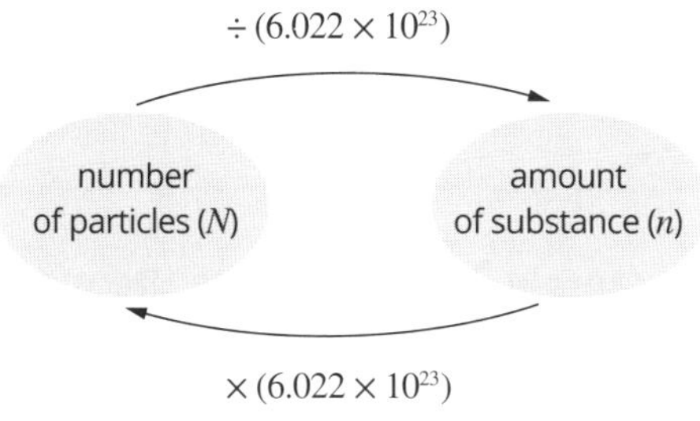

MOLAR MASS

Molar mass is the mass of one mole of an element or compound. For atoms, the molar mass is equal to the relative atomic mass of the element expressed in grams per mole and given the symbol M and the unit $g\,mol^{-1}$. For molecules and compounds, the molar mass is the sum of the relative atomic masses of all of the atoms in the formula (Table 2.2).

TABLE 2.2 Calculation of molar masses of some common elements and compounds

Substance	Relative atomic mass(es)	Molar mass of substance
Na	Na = 22.99	$22.99\,g\,mol^{-1}$
O_2	O = 16.00	$2 \times 16.00 = 32.00\,g\,mol^{-1}$
$NaNO_3$	Na = 22.99 N = 14.01 O = 16.00	$22.99 + 14.01 + (3 \times 16.00)$ $= 85.00\,g\,mol^{-1}$

A useful formula links the amount of a substance (n), its molar mass (M) and the given mass of a substance (m).

$$n = \frac{m}{M}$$

amount in mol → n; m = mass in g; M = molar mass in $g\,mol^{-1}$

PERCENTAGE COMPOSITION AND EMPIRICAL FORMULA

The relative proportions of each element in a compound can be expressed using:

- percentage composition in terms of the mass contributed by each element
- empirical formulae in terms of the number of atoms contributed by each element.

Percentage composition GO TO ➤ page 4

In Module 1, percentage composition by mass was determined using the masses of elements and compounds. The **percentage composition** by mass of elements in a compound can also be determined from its known formula using relative atomic masses:

$$\%\text{ by mass of an element in a compound} = \frac{\text{mass of the element in 1 mol of compound}}{\text{molar mass of the compound}} \times 100$$

Worked example 2.1: Calculate the percentage composition by mass of NH_4NO_3.

Thinking	Working
Calculate %N.	$\%N = \frac{2 \times 14.01}{2 \times 14.01 + 4 \times 1.008 + 3 \times 16.00} \times 100$ $= \frac{28.02}{80.05} \times 100$ $= 35.0\%$
Calculate %H.	$\%H = \frac{4 \times 1.008}{80.05} \times 100$ $= 5.04\%$
Calculate %O.	$\%O = \frac{3 \times 16.00}{80.05} \times 100$ $= 60.0\%$

Empirical formula

Atoms or ions exist in compounds in fixed whole-number ratios. The **empirical formula** of a compound is the simplest whole-number ratio of elements in that compound.

The empirical formula of any compound is determined using the mass of each element present in a given mass of the compound. These masses can be determined experimentally.

ISBN 978 1 4886 1933 5

Worked example 2.2: Determine the empirical formula of a compound, a sample of which contains 25.51% magnesium and 74.49% chloride.

Thinking	Working
Determine the mass of each element present in 100 g of substance from the percentage compositions and write the ratio by mass.	Mg : Cl 25.51 g : 74.49 g
Calculate the amount, in mol, of each element present using $n = \frac{m}{M}$ to determine a mole ratio.	$n(Mg) = \frac{25.51}{24.31}$ = 1.049 mol $n(Cl) = \frac{74.49}{35.45}$ = 2.101 mol
Divide each by the smaller amount to simplify the ratio.	$= \frac{1.049}{1.049}$ = 1 $= \frac{2.101}{1.049}$ = 2
Write the empirical formula.	$MgCl_2$

Molecular formula

The **molecular formula** gives the actual number and type of atoms present in a molecule (Table 2.3). It can be the same as, or different from, the empirical formula.

TABLE 2.3 Some molecular and empirical formulae of organic compounds

Molecule	Molecular formula	Empirical formula
butane	C_4H_{10}	C_2H_5
propanol	C_3H_8O	C_3H_8O
ethanoic acid	$C_2H_4O_2$	CH_2O

The molecular formula is calculated from the empirical formula and molar mass of a compound. It is always a whole-number multiple of the empirical formula.

Worked example 2.3: Determine the molecular formula of a compound with the empirical formula C_2H_5 and a molecular mass of 58.0 g mol^{-1}.

Thinking	Working
Find the molar mass of the empirical formula (EF).	$M(EF) = (2 \times 12.01) + (5 \times 1.008)$ $= 29.06\ g\ mol^{-1}$
Determine the number of empirical formula units in the molecular formula by dividing the molar mass of the molecular formula by the molar mass of the empirical formula.	Number of EF units $= \frac{58.0}{29.06}$ = 2
Write the molecular formula.	$= 2 \times C_2H_5$ $= C_4H_{10}$

CALCULATIONS BASED ON AMOUNT OF REACTANT OR PRODUCT

The ratio provided by the coefficients in a balanced chemical equation is also a **mole ratio** by which atoms or molecules react and are produced. The study of the ratios of moles is called **stoichiometry**.

Consider the reaction represented by the following chemical equation:

$$Mg(s) + 2HCl(aq) \rightarrow MgCl_2(aq) + H_2(g)$$

From this equation it can be seen that:

- 2 mol HCl is required for the complete reaction of 1 mol Mg
- 1 mol Mg and 2 mol HCl will produce 1 mol $MgCl_2$ and 1 mol H_2
- mole ratio Mg : HCl is 1 : 2.

Mass–mass stoichiometry

Determining the mass of product formed from the mass of a reactant requires conversion of mass to mole so the mole ratio can be used (Figure 2.2).

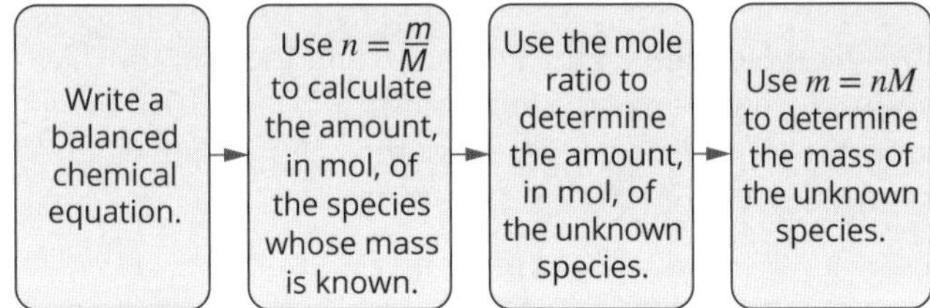

FIGURE 2.2 Flow chart for mass–mass stoichiometric calculations

Worked example 2.4: Determine the mass of H_2 gas formed when 2.43 g Mg reacts with excess HCl.

Thinking	Working
Calculate the amount, in mol, of the given reactant using $n = \frac{m}{M}$.	$n(Mg) = \frac{2.43}{24.31}$ = 0.100 mol
Use the mole ratio from the equation to calculate the amount, in mol, of product.	$\frac{n(Mg)}{n(H_2)} = \frac{1}{1}$ So $n(H_2) = \frac{1}{1} \times n(Mg)$ $= \frac{1}{1} \times 0.100$ = 0.100 mol
Calculate the mass of product using $m = n \times M$.	$m(H_2) = 0.100 \times 2.0$ = 0.20 g

CALCULATIONS BASED ON AMOUNTS OF TWO REACTANTS

When amounts of both reactants are given for a chemical reaction, it must be determined which reactant is completely consumed in the reaction. This reactant is termed the **limiting reactant**. The reactant that is not fully consumed is said to be the **excess reactant**. The amount of the limiting reactant is used in stoichiometric calculations to work out the amount of products formed.

 ISBN 978 1 4886 1933 5

The limiting reactant can be identified using the following steps:

1. Determine the amounts, in mol, of both reactants (say, reactant A and reactant B).
2. Use the amount of reactant A and the mole ratio to determine the amount of reactant B required to react with all of reactant A.
3. If there is not enough of reactant B present, reactant B will be the limiting reactant. If there is more than enough reactant B present, it will be the reactant in excess.

Worked example 2.5: Determine the limiting reactant when 2.43 g Mg reacts with 5.00 g dissolved HCl.

$Mg(s) + 2HCl(aq) \rightarrow MgCl_2(aq) + H_2(g)$

Thinking	Working
Calculate the amount, in mol, of both of the given reactants using $n = \frac{m}{M}$	$n(Mg) = \frac{2.43}{24.31}$ $= 0.100\ mol$ $n(HCl) = \frac{5.00}{36.46}$ $= 0.141\ mol$
Use the mole ratio to calculate the amount of HCl required to react with the total amount of Mg.	$\frac{n(HCl)}{n(Mg)} = \frac{2}{1}$ $n(HCl) = \frac{2}{1} \times n(Mg)$ $n(HCl) = \frac{2}{1} \times 0.100$ $= 0.200\ mol$
Identify the limiting reactant by comparing the amount required to the actual amount.	The amount of HCl required, 0.200 mol, is greater than the actual amount of HCl present, 0.141 mol. Therefore HCl is the limiting reactant.

Concentration and molarity

CONCENTRATION OF SOLUTIONS

Solutes dissolve in solvents to form solutions. Concentration is a measure of the amount of solute dissolved in a particular volume of solution.

Measurement of concentration

The concentration of a solution can also be measured as the mass of solute dissolved per given quantity of solution. A number of different concentration measures are used in various domestic, environmental, commercial and industrial applications (Table 2.4).

TABLE 2.4 Different measures used for expressing concentration of solution

Unit	Description	Example of use
$g\,L^{-1}$	mass of solute, in g, per litre of solution	A bottle of bleach gives the concentration of sodium hypochlorite as $56\,g\,L^{-1}$.
$mg\,L^{-1}$	mass of solute, in mg, per litre of solution	Sydney's water supply has calcium ion concentrations typically around $5–10\,mg\,L^{-1}$.
$mol\,L^{-1}$	moles per litre of solution	Hydrochloric acid used in an analysis in a laboratory is labelled as $1.0\,mol\,L^{-1}$.
%(w/w)	percentage by mass; describes the mass of solute, in g, per 100 g of solution	A bottle of hydrogen peroxide is labelled 10%(w/w).
%(w/v)	percentage mass by volume; describes the mass of solute, in g, per 100 mL of solution	Saline drips are a solution of 0.9%(w/v) sodium chloride.
%(v/v)	percentage by volume; describes the volume of solute, in mL, per 100 mL of solution	Alcohol content of wine might be 13.5%(v/v).
ppm	parts per million, e.g. mass, in g, per million equivalent to $mg\,kg^{-1}$, or $mg\,L^{-1}$	The legal limit of mercury, for safe human consumption, is 0.5 ppm for most species of Australian fish.
ppb	parts per billion, e.g. mass, in g, per billion equivalent to $\mu g\,kg^{-1}$, or $\mu g\,L^{-1}$	The World Health Organization's guideline value for a safe level of arsenic in drinking water is 10 ppb.

Conversion between units

Some useful relationships are:

- $1\,g = 10^3\,mg$
- $1\,mg = 10^3\,\mu g$
- $1\,g = 10^6\,\mu g$
- $1\,ppm = 1\,\mu g\,g^{-1} = 1\,mg\,kg^{-1}$
- $1\,ppb = 1\,\mu g\,kg^{-1}$

MOLAR CONCENTRATION

Chemists often measure concentration in terms of **molarity**, which is the amount, in mol, of solute dissolved per litre of solution ($mol L^{-1}$). Molarity is given the symbol *c*. A solution with a concentration, or molarity, of 5 $mol L^{-1}$ is said to be a '5 molar' solution. A useful formula links the amount of substance (*n*), the concentration of a solution (*c*) and the volume of the solution (*V*).

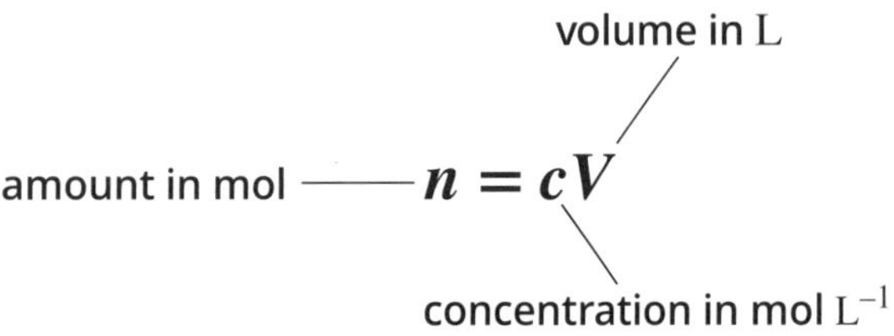

The concentration of the same solution can be expressed in a number of different ways (Figure 2.3).

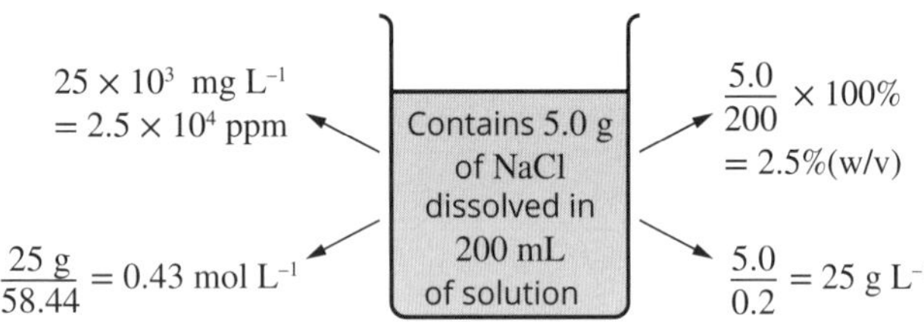

FIGURE 2.3 Different ways of expressing the concentration of a solution

DILUTION

Dilution of a solution by adding water changes the concentration of the solution (Figure 2.4).

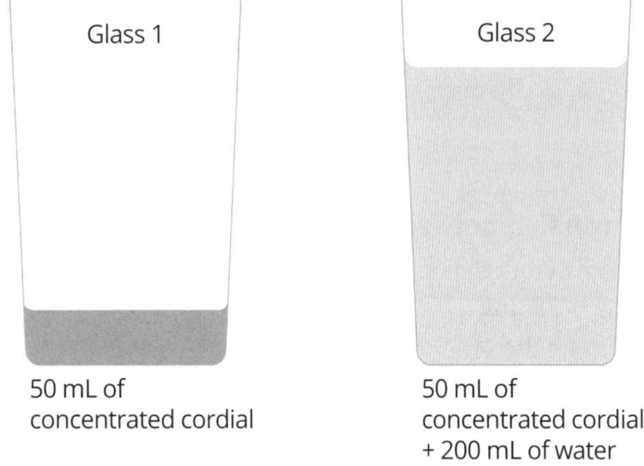

FIGURE 2.4 When cordial is diluted with water the amount of solute does not change. The solute particles, however, are more widely spaced and the molarity has decreased.

The amount of solute does not change, so $n_1 = n_2$. A useful relationship for calculating the concentration of the diluted solution is:

$$c_1V_1 = c_2V_2$$

where c_1 and V_1 are the concentration and volume of the initial solution and c_2 and V_2 are the concentration and volume of the diluted solution.

It is often necessary to dilute a solution (Figure 2.4) before carrying out an analysis that uses stoichiometry in order to obtain convenient volumes. When this occurs, a dilution factor can be considered.

A 25.00 mL sample diluted to 250.0 mL will have a dilution factor of $\frac{250}{25} = 10$. The concentration determined by titration using the diluted sample will need to be multiplied by 10 to gain the undiluted concentration.

STANDARD SOLUTIONS

A standard solution is a solution that has an accurately known concentration. It can be made using a **primary standard** (Figure 2.5). A primary standard is a substance so pure that the amount can be accurately calculated from its mass, for example Na_2CO_3. For a chemical to be a primary standard it should be easily obtained, have a known formula and be easily stored without reacting with water or gases in the atmosphere.

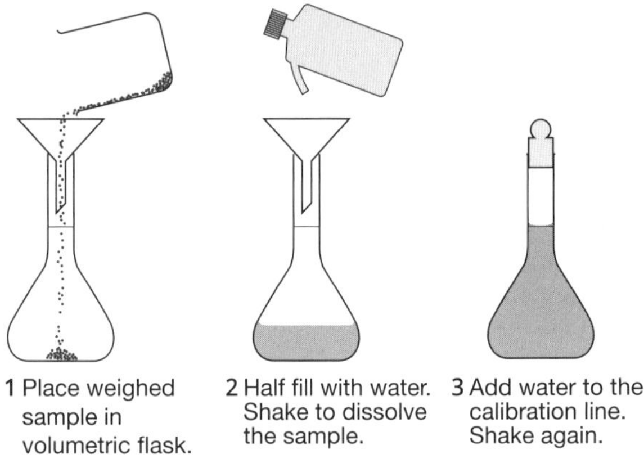

FIGURE 2.5 Standard solution being prepared in a volumetric flask from a primary standard

The concentration, in $mol L^{-1}$, of a standard solution prepared from a primary standard can be calculated by following the steps outlined in Figure 2.6.

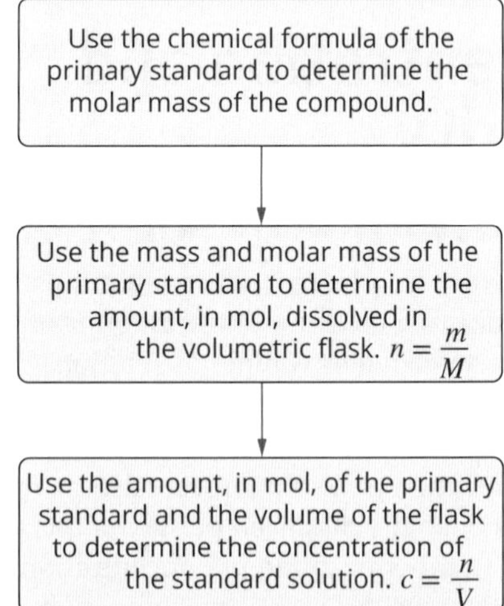

FIGURE 2.6 Flow chart of the steps used in the calculation of the concentration of a standard solution

ISBN 978 1 4886 1933 5

Gas laws

INTRODUCING GASES

Gases share the following physical properties:

- low density
- fill all available space
- easily compressed
- mix rapidly.

The chemical properties of gases depend on the nature of the molecules or atoms in the gas. However, physical properties do not change for different gases. The **kinetic molecular theory** is a model used to explain gas behaviour. Kinetic molecular theory has the following components:

- Gases are composed of small particles (atoms or molecules). Most of the volume of a gas is empty space.
- Gas particles move rapidly in random straight-line motion. They collide with each other and the walls of a container.
- Forces between gas particles are extremely weak. Particles move around virtually independently.
- Collisions between gas particles are elastic, that is, energy is conserved.
- Average kinetic energy of gas particles increases as the temperature of the gas is increased.

Units for gas measurements

Physical properties of gases that can be measured include:

- amount
- volume
- pressure
- temperature.

The amount of a gas is measured in terms of moles where 1 mol of a gas contains Avogadro's constant of gas particles. The symbol n is used for amount (mol).

Volume, symbol V, is the space occupied by a gas. A number of units are commonly used to measure gas volume (Table 2.5).

TABLE 2.5 Example of equivalent volumes

mL	cm³	L	dm³	m³
1.0×10^3	1.0×10^3	1.0	1.0	1.0×10^{-3}

Pressure is given the symbol P. Pressure is defined as the force exerted on a unit area of substance. A number of different units are used to measure pressure (Table 2.6).

TABLE 2.6 Example of equivalent pressures

atm	mmHg	kPa	Pa	bar	$N\,m^{-2}$
1.00	760	101.3	101 325	1.01	101300

In a mixture of gases, each gas exerts a partial pressure. The total pressure of the mixture of gases is the sum of the individual partial pressures.

When used in calculations, temperature must be stated using the Kelvin, rather than the Celsius, scale. The temperature in degrees Celsius (t) can be converted to the temperature in kelvin (T) using the formula $T = t + 273$ (Table 2.7).

TABLE 2.7 Temperatures expressed in degrees Celsius and in kelvin

t (°C)	T (K)
−273	0
0	273
100	373

THE GAS LAWS

The gas laws explain the relationships between different physical properties of gases (Table 2.8).

TABLE 2.8 Summary of the gas laws

Law	Description	Equation
Boyle's law	Volume of a fixed amount of gas at constant temperature is inversely proportional to its pressure.	$P_1V_1 = P_2V_2$ P is pressure (kPa) V is volume (L)
Gay-Lussac's law	Pressure of a fixed amount of gas at fixed volume is directly proportional to the kelvin temperature.	$\frac{P_1}{T_1} = \frac{P_2}{T_2}$ P is pressure (kPa) T is temperature (K)
Charles' law	Volume of a fixed amount of gas at constant pressure is directly proportional to the kelvin temperature.	$\frac{V_1}{T_1} = \frac{V_2}{T_2}$ V is volume (L) T is temperature (K)
Avogadro's law	Volume of a gas at constant pressure and temperature is directly proportional to the amount (mol).	$\frac{V_1}{n_1} = \frac{V_2}{n_2}$ V is volume (L) n is amount (mol)
Combined gas law	Volume of a fixed amount of gas is directly proportional to the kelvin temperature and inversely proportional to the pressure.	$\frac{P_1V_1}{T_1} = \frac{P_2V_2}{T_2}$

THE IDEAL GAS LAW

The ideal gas law is a relationship between the amount, volume, pressure and temperature of a gas.

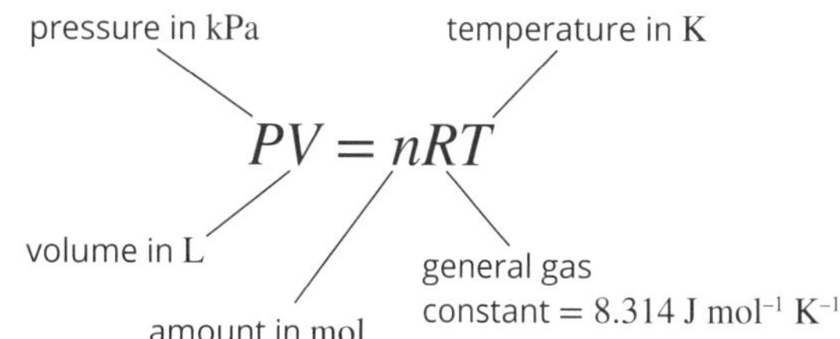

The amount, in mol, of a gas under any particular conditions of volume, pressure and temperature can be calculated by rearranging the ideal gas equation to:

$$PV = nRT$$

Molar volume

The molar volume of a gas is the volume occupied by one mole of the gas. Molar volume has the symbol V_m. The molar volume of different gases will be the same provided the temperature and pressure are the same. Two sets of conditions are used to give a standard value of molar volume (Table 2.9).

TABLE 2.9 Standard conditions for molar volume

Standard condition	V_m (L)	T (°C)	P (kPa)
STP (standard temperature and pressure)	24.8	25	100
SLC (standard laboratory conditions)	22.7	0	100

STOICHIOMETRY INVOLVING GASES AND SOLUTIONS

The principles of stoichiometry can be extended to equations involving gases and solutions in which the known or unknown quantities are other than mass.

In mass–volume stoichiometry, the mass of one reactant or product is determined from the concentration and volume of a solution or the volume of a gas at a particular pressure and temperature. Alternatively, the volume of a gas or volume or concentration of a solution can be determined from mass.

Worked example 2.6: Determine the volume of H_2 gas formed at 35°C and 1.10 atm of pressure when 2.43 g Mg reacts with excess HCl.

Thinking	Working
Calculate the amount, in mol, of the given reactant using $n = \frac{m}{M}$.	$n(\text{Mg}) = \frac{2.43}{24.31}$ $= 0.100$ mol
Use the stoichiometric ratio from the balanced chemical equation to calculate the amount, in mol, of the unknown species.	$n(\text{Mg}) : n(\text{H}_2)$ is 1 : 1 $n(\text{H}_2) = \frac{1}{1} \times n(\text{Mg})$ $= 0.100$ mol
Convert the pressure and temperature to kPa and K.	$P = \frac{1.10}{1.00} \times 101.3$ $= 111$ kPa $T = 35 + 273$ $= 308$ K
Calculate the volume of the gaseous product using the ideal gas law.	$V = \frac{nRT}{p}$ $= \frac{0.100 \times 8.314 \times 308}{111}$ $= 2.31$ L

Consider the reaction represented by the following balanced chemical equation:

$$Mg(s) + 2HCl(aq) \rightarrow MgCl_2(aq) + H_2(g)$$

In volume–volume stoichiometry for gases, the volume does not always need to be converted to number of moles. If all of the reactants and products are in the gaseous state, stoichiometric calculations do not require the calculations of the number of moles. The same number of moles of different gases occupies the same volume at constant temperature and pressure, so the mole ratio in a balanced chemical equation can also be used as a volume ratio.

Summary of quantities used in chemistry

When used in equations, quantities must be expressed in specific units (Table 2.10).

TABLE 2.10 Common quantities and units used in calculations in chemistry

Quantity	Symbol	Unit	Equation(s)
amount	n	mol	$n = \frac{m}{M}$ $n = cV$ $n = \frac{PV}{RT}$ $n = \frac{V}{V_M}$ $\frac{V_1}{n_1} = \frac{V_2}{n_2}$
mass	m	g	$m = n \times M$
molar mass	M	g mol^{-1}	$M = \frac{m}{n}$
volume	V	L	$V = \frac{c}{n}$ $V = n \times V_m$ $V = \frac{nRT}{P}$
concentration	c	mol L^{-1}	$c = \frac{n}{V}$ $c_1V_1 = c_2V_2$
pressure	P	kPa	$P = \frac{nRT}{V}$ $P_1V_1 = P_2V_2$
temperature	T	K	$T = \frac{PV}{nR}$ $\frac{V_1}{T_1} = \frac{V_2}{T_2}$ $\frac{P_1}{T_1} = \frac{P_2}{T_2}$

ISBN 978 1 4886 1933 5

WORKSHEET 2.1

Knowledge review—elements, compounds, atoms and isotopes

1 Mark each statement about elements, compounds, atoms and isotopes as true or false.

Statement	True or false?
An element contains only one or two different types of atom.	
A compound can exist as a molecule or lattice.	
A mixture contains elements and compounds that are not chemically bonded.	
${}^{15}_{7}X$ and ${}^{15}_{8}Y$ are isotopes.	
The relative atomic mass of copper is 63.5.	
The ion ${}^{52}_{24}Cr^{2-}$ has 26 electrons.	
All elements in group 16 of the periodic table have 16 valence electrons.	
Alpha particle radiation consists of high energy electrons.	
The standard mass to which other relative masses are compared is carbon-12, being exactly 10 units.	
The relative atomic mass is a weighted average of the isotopic masses of all of the isotopes that exist for an element.	

2 Write the formula of each of the following compounds and indicate the type of bonding holding the atoms or ions in the compound together.

a magnesium chloride formula: ____________ type of bonding: ____________

b sodium nitride formula: ____________ type of bonding: ____________

c sulfur trioxide formula: ____________ type of bonding: ____________

d dinitrogen tetraoxide formula: ____________ type of bonding: ____________

e aluminium oxide formula: ____________ type of bonding: ____________

f carbon monoxide formula: ____________ type of bonding: ____________

g hydrogen chloride formula: ____________ type of bonding: ____________

h potassium sulfate formula: ____________ type of bonding: ____________

i sodium nitrate formula: ____________ type of bonding: ____________

j carbon dioxide formula: ____________ type of bonding: ____________

WORKSHEET 2.2

Maintaining balance—chemical equations and stoichiometric ratios

1 Balance each chemical equation by adding a coefficient before each reactant and product.

a $\_\_H_2(g) + \_\_O_2(g) \rightarrow \_\_H_2O(l)$

b $\_\_Al(s) + \_\_F_2(g) \rightarrow \_\_AlF_3(s)$

c $\_\_H_2SO_4(aq) + \_\_Mg(OH)_2(aq) \rightarrow \_\_MgSO_4(aq) + \_\_H_2O(l)$

d $\_\_CuCl_2(aq) + \_\_AgNO_3(aq) \rightarrow \_\_AgCl(s) + \_\_Cu(NO_3)_2(aq)$

e $\_\_HCl(aq) + \_\_Al_2O_3(aq) \rightarrow \_\_AlCl_3(aq) + \_\_H_2O(l)$

2 Write a balanced chemical equation for each of the following reactions.

a Copper metal reacts with silver nitrate solution to produce silver metal and copper nitrate.

____

b Sodium metal and gaseous chlorine react to produce solid sodium chloride.

____

c A mixture of solid carbon and copper(II) oxide is heated to produce solid copper and carbon dioxide.

____

d Solid ammonium carbonate decomposes and produces water, ammonia and carbon dioxide.

____

e Solutions of iron(III) chloride and sodium hydroxide are mixed to produce solid iron(III) hydroxide and sodium chloride solution.

____

3 Identify the stoichiometric ratios in the following balanced chemical equations.

a $2CO(g) + O_2(g) \rightarrow 2CO_2(g)$

Ratio: ____

b $4Al(s) + 3O_2(g) \rightarrow 2Al_2O_3(s)$

Ratio: ____

c $2Na(s) + 2H_2O(l) \rightarrow 2NaOH(aq) + H_2(g)$

Ratio: ____

d $2HCl(aq) + Na_2CO_3(aq) \rightarrow 2NaCl(aq) + H_2O(l) + CO_2(g)$

Ratio: ____

RATING MY LEARNING	My understanding improved	Not confident ◄——► Very confident ○ ○ ○ ○ ○	I answered questions without help	Not confident ◄——► Very confident ○ ○ ○ ○ ○	I corrected my errors without help	Not confident ◄——► Very confident ○ ○ ○ ○ ○

ISBN 978 1 4886 1933 5

WORKSHEET 2.3

Marvellous moles—the chemist's unit of measurement

1 The mole is a unit of measurement. It represents a physical quantity in much the same way as a more common unit of measurement, the dozen.

a Compare the use of the two units by calculating the actual numbers of the following amounts:

1 dozen = ______________________ 1 mol = ______________________

4 dozen = 4 × 12 = ______________________ 4 mol = $4 \times 6.022 \times 10^{23}$ = ______________________

0.10 dozen = ______________________ 0.10 mol = ______________________

b Describe one similarity and one difference between the units 'mole' and 'dozen'.

____

____

2 Explain why a unit of measurement such as the mole is necessary in chemistry.

____

____

3 Examine the diagram, which shows a sample of common table sugar, sucrose ($C_{12}H_{22}O_{11}$), on a balance, and its mass in grams. The relative atomic masses of the elements found in sucrose are shown in the table.

Relative atomic masses of the elements found in sucrose	
Element	A_r
H	1.008
C	12.01
O	15.99

Complete the following table by first identifying the formula needed for each calculation and then solving for the answer. Don't forget to include the correct units in your answers. The first few formulae have been included for you.

Calculation	Relevant formula	Solution
molar mass of sucrose	$M(C_{12}H_{22}O_{11})$ = sum of relative atomic masses of elements in compound	
amount, in mol, of sucrose in sample	$n(C_{12}H_{22}O_{11}) = \frac{m}{M}$	
number of sucrose particles in sample	$N(C_{12}H_{22}O_{11}) = n \times N_A$	
amount, in mol, of carbon atoms in sample	$n(C) = 12 \times n(C_{12}H_{22}O_{11})$	
number of carbon atoms in sample		
total amount, in mol, of atoms in sample		
total number of atoms in sample		

RATING MY LEARNING	My understanding improved	Not confident ◄—► Very confident ○ ○ ○ ○ ○	I answered questions without help	Not confident ◄—► Very confident ○ ○ ○ ○ ○	I corrected my errors without help	Not confident ◄—► Very confident ○ ○ ○ ○ ○

WORKSHEET 2.4

Stoichiometry 1—mass–mass calculations

Stoichiometry allows calculation of the amounts of reactants consumed and products formed from a given amount of any one reactant or product. In mass–mass stoichiometry, the mass of a reactant is used to determine the mass of a product.

A 10.0 g sample of magnesium carbonate is added to an excess amount of hydrochloric acid and allowed to completely react.

1 Write a balanced chemical equation for the reaction between $MgCO_3$ and HCl. Products of the chemical reaction are magnesium chloride, water and carbon dioxide gas.

____

2 Calculate the mass of carbon dioxide formed from the mass of magnesium carbonate reacted by completing the following steps.

a Calculate the amount, in mol, of $MgCO_3$.

____

____

b Identify the mole ratio of $MgCO_3$ reacted to CO_2 produced as per the chemical equation.

____

c Use the mole ratio to calculate the amount, in mol, of CO_2 produced.

____

d Calculate the mass of CO_2 produced.

____

____

3 The other reactant in this reaction is hydrochloric acid. What is meant by describing it as 'in excess'?

____

____

4 Use a similar series of steps to those outlined in question **2** to calculate the mass of magnesium carbonate required to react to produce 10.0 g CO_2.

____

____

____

____

____

____

____

RATING MY LEARNING	My understanding improved	Not confident ◄——► Very confident ○ ○ ○ ○ ○	I answered questions without help	Not confident ◄——► Very confident ○ ○ ○ ○ ○	I corrected my errors without help	Not confident ◄——► Very confident ○ ○ ○ ○ ○

 ISBN 978 1 4886 1933 5

WORKSHEET 2.5

Molarity—measuring moles in solution

1 A number of similar terms are used when dealing with quantities in chemistry. Outline the use of each of the following terms in the space provided.

Mole: ______________________________

____

Molar mass: ______________________________

____

Molarity: ______________________________

____

2 Each beaker shown contains a solution of potassium chloride. Consider the diagram and answer the following questions.

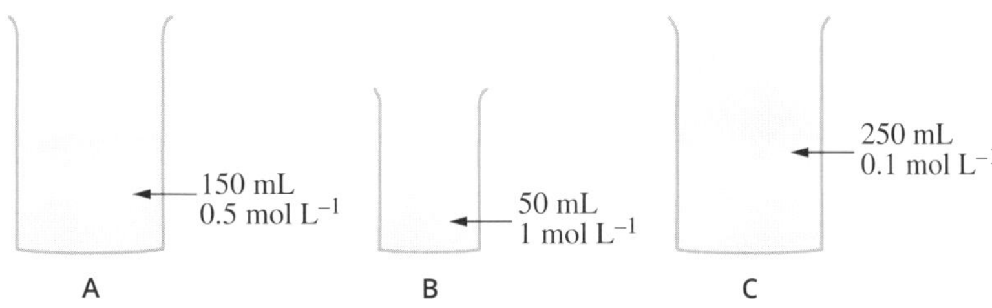

a Which beaker has the highest molarity—A, B or C? ____________

b Calculate the amount, in mol, of potassium chloride dissolved in each beaker.

____

____

____

c Does the beaker with the highest molarity also contain the greatest mass of dissolved solute? Explain your answer.

____

____

____

d A student dilutes the solution in beaker A by adding water until it holds the same volume of solution as beaker C. Calculate the new molarity of the solution in beaker A.

____

____

3 For each solution, circle the concentrations that are equivalent to the concentration in the first column.

300 mg L^{-1} NaCl solution	0.300 g L^{-1}	3.00 g L^{-1}	0.0051 mol L^{-1}
25 ppm Mg^{2+} in water	25 mg L^{-1}	25 g L^{-1}	25 μg g^{-1}
12.5% (w/v) ethanol in wine	12.5 g L^{-1}	125 g L^{-1}	1.25×10^6 ppm
2.0 mol L^{-1} NaOH	0.050 g L^{-1}	8.0×10^4 mg L^{-1}	80 ppm

RATING MY LEARNING	My understanding improved	Not confident ◄——► Very confident ○ ○ ○ ○ ○	I answered questions without help	Not confident ◄——► Very confident ○ ○ ○ ○ ○	I corrected my errors without help	Not confident ◄——► Very confident ○ ○ ○ ○ ○

WORKSHEET 2.6

Gases—the ideal gas equation

The ideal gas equation combines amount, pressure, volume and temperature and several gas laws. Answer the questions about each of these concepts.

1 The following measurements of pressures are equivalent:

1.00 atm = 760 mmHg = 101.3 kPa = 1.013 bar

100 kPa = 1.00 bar

Convert the pressures shown to the alternative units in each case. An example is shown.

1.30 atm to mmHg 1.00 atm = 760 mmHg So 1.30 atm = 1.30 × 760 = 988 mmHg	145 kPa to bar	3.50 bar to atm
480 mmHg to kPa	1.50 atm to kPa	150 mmHg to bar

2 The following measurements of volume are equivalent:

1.0×10^3 mL = 1.0×10^3 cm^3 = 1.0 L = 1.0 dm^3 = 1.0×10^{-3} m^3

Convert the volumes shown to the alternative units in each case. An example is shown.

1.3 dm^3 to mL 1 dm^3 = 1000 mL So 1.3 dm^3 = 1.3 × 1000 = 1.3×10^3 mL	300 mL to L	4.3 L to m^3
2.8 L to mL	0.85 m^3 to L	480 cm^3 to dm^3

3 The gas laws specify relationships between two physical properties of gases. For each question, identify the relevant law that must be used to calculate the quantity, then carry out the calculation.

a A fixed amount of gas occupies a volume of 2.3 L at 25°C. The temperature is increased to 50°C. Calculate the new volume of the gas if the pressure is kept constant.

Relevant gas law: ______________________

Calculation:

____

____

____

b 6.0 mol of a gas occupies a volume of 5.0 L. Calculate the new volume of the gas if 3.5 mol of gas is added while keeping the temperature and pressure constant.

Relevant gas law: ______________________

Calculation:

____

____

____

ISBN 978 1 4886 1933 5

c A fixed amount of gas has a pressure of 120 kPa at 0°C. The temperature is increased to 100°C. Calculate the new pressure of the gas if the volume is kept constant.

Relevant gas law: ______________________

Calculation:

____

____

____

4 The ideal gas equation, $PV = nRT$, combines the results of different scientists' experiments on gases into a single equation. The value of the ideal gas constant, R, in the equation depends on the units used for pressure, temperature and volume.

If the value of $R = 8.314\ \text{J mol}^{-1}\,\text{K}^{-1}$ in the ideal gas equation, complete the following statements for the units used for pressure, volume and temperature.

Pressure in ____________

Volume in ____________

Temperature in ____________

5 Methane, CH_4, is the major component of natural gas. It is used for heating and cooking in homes.

a When methane undergoes combustion it reacts with oxygen gas to produce carbon dioxide gas and water vapour. Write a balanced chemical equation for this reaction.

____

____

b Use stoichiometry and the ideal gas equation to calculate the volume of the greenhouse gas CO_2 (measured at 25°C and 101.3 kPa) that would be produced when 1.00 kg of methane undergoes combustion.

____

____

____

____

____

____

RATING MY LEARNING	My understanding improved	Not confident ◄—► Very confident ○ ○ ○ ○ ○	I answered questions without help	Not confident ◄—► Very confident ○ ○ ○ ○ ○	I corrected my errors without help	Not confident ◄—► Very confident ○ ○ ○ ○ ○

WORKSHEET 2.7

Stoichiometry 2—mass–volume calculations

A three-step process is generally at the heart of stoichiometric calculations:

i The amount, in mol, of the given limiting quantity is calculated using an appropriate relationship, usually one of:

$n = \frac{m}{M}$ $\quad n = cV$ $\quad n = \frac{V}{V_M}$ $\quad n = \frac{PV}{RT}$

ii The mole ratio provided in a balanced chemical equation is used to calculate the amount, in mol, of the unknown quantity from the amount, in mol, of the given limiting quantity.

iii The desired quantity is calculated using a rearrangement of the appropriate relationship.

Consider the reaction between solid sodium carbonate and hydrochloric acid.

1 Balance the chemical equation by adding coefficients in front of the appropriate reactants and products.

$\_\_Na_2CO_3(s) + \_\_HCl(g) \rightarrow \_\_NaCl(aq) + \_\_H_2O(l) + \_\_CO_2(g)$

2 4.50 g of Na_2CO_3 is reacted with an excess amount of HCl. Use the three-step process to calculate:

a volume of 0.100 mol L^{-1} HCl required to react completely with the sodium carbonate

b mass of NaCl produced

c mass of CO_2 produced

d volume of CO_2 produced, measured at SLC

e volume of CO_2 produced, measured at 100°C and 110 kPa.

RATING MY LEARNING	My understanding improved	Not confident ◄—► Very confident ○ ○ ○ ○ ○	I answered questions without help	Not confident ◄—► Very confident ○ ○ ○ ○ ○	I corrected my errors without help	Not confident ◄—► Very confident ○ ○ ○ ○ ○

ISBN 978 1 4886 1933 5

WORKSHEET 2.8

Solving complex calculations—using more than one formula

Many calculations in chemistry require a number of different steps. Each of the following problems requires the use of at least two of the formulae shown.

$n = \frac{m}{M}$	$n = cV$	$n = \frac{N}{N_A}$
$PV = nRT$	$c_1V_1 = c_2V_2$	$n = \frac{V}{V_M}$

For each problem:

- plan your method of solution by deciding which of the formulae you will need to use and in what order
- solve the problem, remembering to show your answer with the correct number of significant figures.

The first problem has been started for you.

1 Calculate the mass, in grams, of solute dissolved in 700 mL of 0.10 mol L^{-1} NaOH.

Plan: $n(\text{NaOH}) = cV$, then $m(\text{NaOH}) = nM$

Solve:

2 Calculate the number of HCl molecules present in 2.00 L HCl gas at 20°C and 150 kPa.

Plan:

Solve:

3 2.0 g NaCl is dissolved in water and made up to 100 mL of solution. 10 mL of this solution is then diluted to 250 mL. Calculate the concentration, in mol L^{-1}, of the diluted solution.

Plan:

Solve:

RATING MY LEARNING	My understanding improved	Not confident ◄—► Very confident ○ ○ ○ ○ ○	I answered questions without help	Not confident ◄—► Very confident ○ ○ ○ ○ ○	I corrected my errors without help	Not confident ◄—► Very confident ○ ○ ○ ○ ○

WORKSHEET 2.9

Literacy review—key terms and definitions

The table contains some descriptions of quantities measured in chemistry. Complete the table by writing the correct name of each quantity beside each description.

Description	Quantity
space occupied by a gas or solution	
force per unit surface area exerted by gas particles colliding with the walls of a container	
quantity of substance counted in lots of Avogadro's constant	
example of this quantity is 60 K	
mass of one mole of a substance	
volume of one mole of a gas	
amount of substance dissolved per litre of solution	

The table contains some formulae used in calculations of quantities. Write an explanation for each formula in the space beside each formula.

Formula	Explanation
$n = \frac{m}{M}$	
$c = \frac{n}{V}$	
$c_1V_1 = c_2V_2$	
$\frac{P_1}{T_1} = \frac{P_2}{T_2}$	
$PV = nRT$	
$n = \frac{N}{N_A}$	
$V = n \times VM$	

RATING MY LEARNING	My understanding improved	Not confident ⟷ Very confident ○ ○ ○ ○ ○	I answered questions without help	Not confident ⟷ Very confident ○ ○ ○ ○ ○	I corrected my errors without help	Not confident ⟷ Very confident ○ ○ ○ ○ ○

 ISBN 978 1 4886 1933 5

WORKSHEET 2.10

Thinking about my learning

On completion of Module 2: Introduction to quantitative chemistry, you should be able to describe, explain and apply the relevant scientific ideas. You should be able to interpret, analyse and evaluate data.

1 The table shows the areas of key knowledge covered in this module. Reflect on how well you understand each section. Rate your learning by shading the circle that corresponds to your level of understanding for each area. It may be helpful to use colour as a visual representation. For example:

- green—very confident
- orange—in the middle
- red—starting to develop.

Section focus	**Rate my learning**				
	Starting to develop ⟷ Very confident				
Chemical reactions—writing and balancing chemical equations	○	○	○	○	○
The mole concept—Avogadro's constant, $n = \frac{m}{M}$, molar mass	○	○	○	○	○
Percentage composition and empirical and molecular formulae	○	○	○	○	○
Stoichiometry—calculations based on amount of reactant or product, limiting and excess reactants	○	○	○	○	○
Concentration—different units of concentration, molarity, $n = cV$, dilution, standard solutions	○	○	○	○	○
Gases—kinetic molecular theory, pressure, volume, amount, temperature and their units	○	○	○	○	○
Gas laws—Boyle's, Gay-Lussac's, Charles', Avogadro's, combined and ideal gas law.	○	○	○	○	○
Stoichiometry involving gases and solutions	○	○	○	○	○

2 Consider points you have shaded from 'starting to develop' to 'in the middle'. List specific ideas you can identify that were challenging.

____

____

____

____

____

____

3 Write down two different strategies that you will apply to help further your understanding of these ideas.

____

____

____

____

____

PRACTICAL ACTIVITY 2.1

Molar mass of an element and a compound

Suggested duration:
15 minutes for part A plus several hours standing time and drying time overnight
40 minutes for part B plus drying time overnight

INTRODUCTION

In part A of this practical activity, you will determine the number of moles of copper produced in a reaction between copper oxide and zinc metal. From measurements of the mass of the copper, the molar mass of the element copper can be calculated.

In part B, you will determine the number of moles of barium sulfate produced in a precipitation reaction between sodium sulfate and barium chloride. From measurements of the mass of the dried barium sulfate, the molar mass of the compound barium sulfate can be calculated.

PURPOSE

To determine the molar mass of an element, copper, and a compound, barium sulfate.

MATERIALS

- 2 g solid copper oxide (black), CuO
- 3 g zinc granules or powder
- 50 mL 2 mol L^{-1} sulfuric acid, H_2SO_4
- 250 mL beaker
- weighing bottle
- evaporating basin
- electronic balance
- Bunsen burner, tripod stand and gauze mat
- safety glasses
- 50 mL measuring cylinder
- stirring rod
- spatula

Part A—Molar mass of copper

As directed by your teacher, complete the risk assessment and management table by referring to the hazard labels on the reagent bottles or safety data sheets (SDS) or your teacher's risk assessment for the activity.

PRE-LAB SAFETY INFORMATION		
Material used	**Hazard**	**Control**
copper oxide		
zinc granules		
2 mol L^{-1} H_2SO_4		
zinc oxide		

Please indicate that you have understood the information in the safety table.

Name (print): ______________________________

I understand the safety information (signature): ______________________________

PROCEDURE

Record all of your procedure results in Table 1.

1 Using a weighing bottle, find the exact mass of approximately 2 g CuO.

2 Transfer the copper oxide to a 250 mL beaker. Add 50 mL of 2 mol L^{-1} H_2SO_4. If necessary, heat the mixture by standing the 250 mL beaker in a large beaker of hot water. Stir with a glass rod until the copper oxide dissolves.

3 When all the copper oxide has reacted, add 50 mL water and approximately 3 g granulated zinc. Swirl and allow to react overnight.

4 If the mixture is still blue at the next lesson, add another 1 g granulated zinc and allow to stand for several hours.

5 Decant the supernatant liquid. Carefully wash the deposited copper with about 50 mL water. Wash it three more times.

6 Find the mass of a clean, dry evaporating basin.

7 Using the minimum amount of water, transfer the copper from the beaker to the evaporating basin.

8 Dry the copper by heating gently over a Bunsen burner or similar. When cool, find the mass of the evaporating basin plus copper.

ISBN 978 1 4886 1933 5

RESULTS

TABLE 1 Mass results

Substance	Mass (g)
copper oxide	
evaporating basin	
evaporating basin and copper	

PROCESSING DATA

1 Calculate the mass of copper that was in the copper oxide.

2 Write a balanced equation for the reaction between copper oxide and zinc metal. The products are zinc oxide and copper metal. You do not need to include states.

3 Use your starting mass of copper oxide to calculate the amount, in mol, of copper oxide reacted.

4 Using the amount of copper oxide reacted and the mole ratio in the balanced equation, calculate the amount, in mol, of copper produced.

5 Use the amount, in mol, and the mass of copper to calculate a molar mass of copper.

ANALYSIS OF RESULTS

6 According to the periodic table, the molar mass of copper gas is 63.55 g mol^{-1}. How close was your experimentally determined value? Suggest two sources of experimental error that may have occurred during your experiment.

Part B—Molar mass of barium sulfate

As directed by your teacher, complete the pre-lab safety information by referring to the hazard labels on the reagent bottles or safety data sheets (SDS) or your teacher's risk assessment for the activity.

PRE-LAB SAFETY INFORMATION		
Material used	**Hazard**	**Control**
solid sodium sulfate		
0.5 mol L^{-1} barium chloride solution		
2 mol L^{-1} hydrochloric acid		
barium sulfate		
heating of solution and crucible	burning and scalding	
Please indicate that you have understood the information in the safety table. Name (print): ______ I understand the safety information (signature): ______		

MATERIALS

- 0.5 g solid sodium sulfate, Na_2SO_4
- 20 mL 0.5 mol L^{-1} barium chloride solution, $BaCl_2$
- 3 mL 2 mol L^{-1} hydrochloric acid, HCl
- 200 mL de-ionised water
- 20 mL warm de-ionised water
- 2 × 100 mL beakers
- 600 mL beaker
- 10 mL measuring cylinder
- burette and stand
- filter funnel
- filter paper
- vacuum flask and vacuum pump (water-jet type)
- glass filter crucible (No. 4 porosity) and rubber adaptor*
- stirring rod
- wash bottle containing de-ionised water
- Bunsen burner, tripod stand and gauze mat
- bench mat
- electronic balance
- oven
- safety glasses

* A Gooch crucible, rubber adaptor and filter paper to suit may be used instead. The Gooch crucible is less expensive, but filtration often takes longer and uses much more water.

PROCEDURE

Record all of your procedure results in Table 2.

1 Weigh out accurately about 0.5 g of sodium sulfate.

2 Add the Na_2SO_4 to 50 mL de-ionised water in a 600 mL beaker and stir to dissolve as much of the sample as possible.

3 Add about 3 mL of 2 mol L^{-1} HCl to the solution of Na_2SO_4 and add more water so that the total volume is about 200 mL. Heat the solution until it boils.

4 Add 15 mL of 0.5 mol L^{-1} $BaCl_2$ solution drop by drop from a burette to the hot solution, stirring continuously. A white precipitate of barium sulfate will form.

5 Boil the mixture for a further minute, then remove it from the heat and allow the precipitate to settle. Ensure no sulfate ions remain in the solution by adding several more drops of $BaCl_2$ solution. If more precipitate forms, add 3 mL $BaCl_2$ solution and test again for unreacted sulfate ions. If desired, the mixture can be left to stand overnight at this stage.

6 Weigh a glass filter crucible.

7 Collect the precipitate in the glass filter crucible using gentle vacuum filtration. (Filtration is faster if most of the liquid is filtered before the bulk of the precipitate is collected in the crucible.)

 ISBN 978 1 4886 1933 5

8 Use about 10 mL warm de-ionised water to wash any precipitate remaining in the beaker into the crucible.

9 Place the crucible and contents in an oven heated to 100–110°C and leave overnight.

10 Weigh the crucible and contents and record the mass in Table 2.

RESULTS

TABLE 2 Mass results

Substance	Mass (g)
sodium sulfate	
crucible	
crucible and barium sulfate	

PROCESSING DATA

1 Calculate the mass of the barium sulfate precipitate collected.

2 Write a balanced equation for the reaction between barium chloride and sodium sulfate. The products are barium sulfate and sodium chloride.

3 Calculate the amount, in mol, of sodium sulfate reacted.

4 Use the amount of sodium sulfate reacted and the mole ratio in the balanced equation to calculate the amount, in mol, of barium sulfate produced.

5 Use the amount, in mol, and the mass of barium sulfate to calculate the molar mass of barium sulfate.

ANALYSIS OF RESULTS

6 Use a periodic table to determine the theoretical molar mass of barium sulfate. Comment on how close your experimental result is to the theoretical result.

7 How would the results obtained be affected if the mixture in step 5 of the procedure was not tested with more barium chloride solution?

8 Comment on the reliability of conclusions you can draw from this investigation.

CONCLUSION

RATING MY LEARNING	My understanding improved	Not confident ◄──► Very confident ○ ○ ○ ○ ○	I answered questions without help	Not confident ◄──► Very confident ○ ○ ○ ○ ○	I corrected my errors without help	Not confident ◄──► Very confident ○ ○ ○ ○ ○

 ISBN 978 1 4886 1933 5

PRACTICAL ACTIVITY 2.2

Conservation of mass and determination of empirical formula

Suggested duration: 40 minutes

INTRODUCTION

Elements always combine in a definite mass ratio. When solid zinc and solid iodine are mixed, the rate of reaction is too slow to observe any physical changes. However, if the iodine is first dissolved in alcohol, the rate is very fast and heat energy is released. The alcohol also dissolves the formed compound, zinc iodide, so it can be easily collected. In this investigation, all iodine will be used up but some zinc will remain unreacted.

PURPOSE

To investigate the law of conservation of mass using the masses of reactants and products and to derive the empirical formula of zinc oxide.

PRE-LAB SAFETY INFORMATION

Material used	Hazard	Control
iodine	toxic by all routes of exposure; stains skin and clothing; vapour highly irritating to lungs	Wear eye and skin protection.
ethanol	highly flammable	Keep away from naked flame; use eye protection.
zinc	toxic to aquatic life	Wear safety glasses, gloves and a laboratory coat. Use appropriate disposal procedures.
Product	**Hazard**	**Control**
zinc iodide	moderately toxic if ingested; corrosive	Wash hands after touching it.

Please indicate that you have understood the information in the safety table.

Name (print): ______________________________

I understand the safety information (signature): ______________________________

MATERIALS

- 150 mL conical flask with stopper
- balance
- 1–2 g of small zinc pieces (cut up cleanly polished zinc sheet)
- specimen tube with lid containing 0.8–1.0 g solid iodine
- 20 mL ethanol
- dropping pipette
- evaporating basin
- hotplate
- safety glasses

PROCEDURE

Record all of your procedure results in Table 1.

1. Weigh a small conical flask and stopper to the nearest 0.01 g. Add about 2 g zinc pieces and reweigh as accurately as possible.
2. Transfer all iodine from the specimen tube directly into the conical flask and reweigh the flask with stopper as accurately as possible. Do not worry if some iodine was left behind in the specimen tube.
3. Cautiously add 10 drops of ethanol to the mixture. Replace the stopper and rapidly swirl the flask from side to side to mix the contents. Make sure that the iodine does not touch the stopper.
4. Let the flask cool before adding another 10 drops of ethanol. Continue with the addition of ethanol, 10 drops at a time, until the iodine has turned a pale straw colour. At the end of the reaction, the flask should not contain more than 5 mL ethanol and some unreacted zinc.
5. Carefully weigh a clean, dry evaporating basin. Decant into it the liquid from the flask, making sure that no zinc is transferred into the evaporating basin.

ISBN 978 1 4886 1933 5

6 Rinse the flask three times with 5 mL ethanol, each time rinsing the sides of the flask and decanting the liquid into the evaporating basin.

7 Place the flask on the hotplate, heating gently. Lightly shake the flask constantly as the ethanol evaporates. Stop when the flask is dry and clean, and the zinc pieces are loose.

8 Allow the flask to cool then reweigh the flask, zinc and stopper to the nearest 0.01 g.

9 Heat the evaporating basin strongly on the hotplate until all the ethanol has evaporated and a crust forms. If there is still some liquid trapped under the crust, see your teacher.

10 Allow the basin to cool before reweighing it.

RESULTS

TABLE 1 Mass results

Material	Mass (g)
conical flask	
conical flask with zinc	
conical flask with zinc and iodine	
conical flask and leftover zinc	
evaporating basin	
evaporating basin and zinc iodide	

PROCESSING DATA

1 Calculate the mass of zinc that reacted.

2 Calculate the mass of iodine that reacted.

3 Calculate the mass of zinc iodide that formed.

4 Calculate the:

a amount, in mol, of zinc atoms reacted

b amount, in mol, of iodine atoms reacted

c whole-number ratio of zinc atoms to iodine atoms.

 ISBN 978 1 4886 1933 5

5 From your answer to question **4**, derive the empirical formula of zinc iodide.

6 Write the chemical equation for the reaction performed in this experiment.

ANALYSIS OF RESULTS

7 Do your results support the law of conservation of mass? Account for any differences.

CONCLUSION

RATING MY LEARNING	My understanding improved	Not confident ⟷ Very confident ○ ○ ○ ○ ○	I answered questions without help	Not confident ⟷ Very confident ○ ○ ○ ○ ○	I corrected my errors without help	Not confident ⟷ Very confident ○ ○ ○ ○ ○

PRACTICAL ACTIVITY 2.3

Products of a decomposition reaction

Suggested duration: 40 minutes

INTRODUCTION

Sodium hydrogen carbonate ($NaHCO_3$) decomposes on heating. We can predict that the product of this decomposition will be either sodium oxide (Na_2O) or sodium carbonate (Na_2CO_3). Carbon dioxide (CO_2) gas is also produced.

A balanced equation can be written for each of the predictions. From the mole ratios in each possible equation, the mass of the product that would be formed in each reaction can be calculated. By comparing this with the actual mass obtained in the experiment, the product of the decomposition can be identified.

PURPOSE

To demonstrate that chemicals react in simple whole-number ratios by moles and use mass–mass stoichiometry to identify the products of a decomposition reaction.

MATERIALS

- solid sodium hydrogen carbonate, $NaHCO_3$ (about 5 g)
- crucible and lid
- pipeclay triangle
- Bunsen burner
- bench mat
- tripod
- gauze mat
- tongs
- electronic balance
- safety glasses

PRE-LAB SAFETY INFORMATION

Material used	Hazard	Control
sodium hydrogen carbonate	slightly toxic if ingested	Wear safety glasses, gloves and a laboratory coat.

Please indicate that you have understood the information in the safety table.

Name (print): ______________________________

I understand the safety information (signature): ______________________________

PROCEDURE

Record all of your procedure results in Table 1.

1 Accurately weigh a crucible. Add approximately 5 g of $NaHCO_3$ and reweigh.

2 Heat the crucible for 5 minutes, gently at first and then more strongly. Allow to cool and weigh again.

3 Heat again for a few minutes. Allow to cool and reweigh.

4 Repeat step 3 until no further mass loss occurs.

RESULTS

TABLE 1 Mass results

Material	Mass (g)
crucible	
crucible and sodium hydrogen carbonate	
crucible and contents after first heating	
crucible and contents after second heating	
crucible and contents after third heating	

ISBN 978 1 4886 1933 5

PRACTICAL ACTIVITY 2.3

PROCESSING DATA

1 Two balanced chemical equations for two possible decomposition reactions are given. In each case, calculate the theoretical mass of the solid product from your starting mass of sodium hydrogen carbonate.

a $2NaHCO_3(s) \rightarrow Na_2O(s) + 2CO_2(g) + H_2O(g)$

b $2NaHCO_3(s) \rightarrow Na_2CO_3(s) + CO_2(g) + H_2O(g)$

2 Compare the actual mass of solid product formed in the experiment with the theoretical masses calculated and identify the solid product of the decomposition.

3 Calculate the mass of carbon dioxide produced in the reaction.

DISCUSSION

4 Explain why sodium hydrogen carbonate (bicarbonate of soda) is a useful ingredient in baking.

CONCLUSION

RATING MY LEARNING	My understanding improved	Not confident 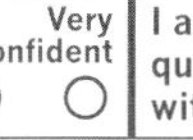 Very confident ○ ○ ○ ○ ○	I answered questions without help	Not confident 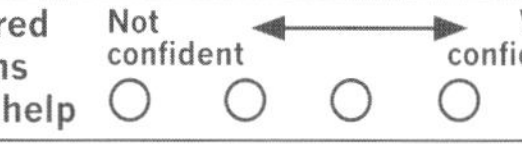 Very confident ○ ○ ○ ○ ○	I corrected my errors without help	Not confident 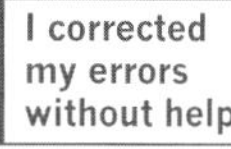 Very confident ○ ○ ○ ○ ○

ISBN 978 1 4886 1933 5

PRACTICAL ACTIVITY 2.4

Preparation of a standard solution

Suggested duration: 20 minutes

INTRODUCTION

Standard solutions have accurately known concentrations and are used in volumetric analysis to determine the concentration of a solution of unknown concentration. Standard solutions can be made by dissolving a measured mass of a primary standard in a known volume of solution.

PURPOSE

To prepare a standard solution from a primary standard.

PRE-LAB SAFETY INFORMATION

Material used	Hazard	Control
sodium carbonate	serious eye irritant; skin irritant	Do not breathe the dust. Wear safety glasses, gloves and a laboratory coat.

Please indicate that you have understood the information in the safety table.

Name (print): ____________________

I understand the safety information (signature): ____________________

MATERIALS

- anhydrous sodium carbonate, Na_2CO_3
- 250 mL de-ionised water
- 250 mL volumetric flask
- weighing bottle or watch glass
- small funnel
- spatula
- dropping pipette
- wash bottle containing de-ionised water
- electronic balance
- paper label or marking pen
- safety glasses

PRE-LABORATORY EXERCISE

Calculate the mass of anhydrous sodium carbonate, Na_2CO_3, required to make 250 mL of 0.100 mol L^{-1} solution.

____

____

____

____

PROCEDURE

1 Weigh out your calculated mass of Na_2CO_3 (within 0.1 g) in a weighing bottle. Record the exact mass of solid used.

2 Use a small funnel to transfer the Na_2CO_3, a little at a time, to a 250 mL volumetric flask. Use a wash bottle of de-ionised water to wash traces of Na_2CO_3 in the weighing bottle and funnel into the flask.

3 Add de-ionised water to the flask until it is almost half full. Stopper and swirl the contents of the flask to dissolve the Na_2CO_3.

4 Add more water until the meniscus of the solution is nearly level with the calibration line. Use a dropping pipette to add the final few drops of de-ionised water so that the bottom of the meniscus is exactly level with the calibration line. (Ensure the flask is standing on a horizontal surface and your eye is at the same level as the line.)

5 Shake the flask by repeatedly inverting it so that the concentration of the solution is uniform. Label the flask with its contents and concentration and your name.

RESULTS

Mass of anhydrous sodium carbonate: __________ g

ISBN 978 1 4886 1933 5

PRACTICAL ACTIVITY 2.4

PROCESSING DATA

1 Calculate the concentration, in mol L^{-1}, of the Na_2CO_3 solution.

2 Convert the concentration you calculated in mol L^{-1} in question **1** to g L^{-1} and ppm.

ANALYSIS OF RESULTS

3 List the properties that make sodium carbonate a good primary standard.

4 Identify some of the sources of error associated with this experiment.

5 How would the concentration of the standard solution be affected if the volumetric flask had been rinsed with de-ionised water before use and droplets of water were left in the flask when the sodium carbonate was added?

CONCLUSION

RATING MY LEARNING	My understanding improved	Not confident ◄—► Very confident ○ ○ ○ ○ ○	I answered questions without help	Not confident ◄—► Very confident ○ ○ ○ ○ ○	I corrected my errors without help	Not confident ◄—► Very confident ○ ○ ○ ○ ○

PRACTICAL ACTIVITY 2.5

Determination of HCl content in brick cleaner

Suggested duration: 40 minutes

INTRODUCTION

Brick cleaner contains concentrated hydrochloric acid as the active ingredient. The acid reacts with the basic components of concrete, enabling concrete to be removed from brickwork. To analyse brick cleaner, a sample is diluted (because the original acid is highly concentrated) and titrated against the standard solution of a base, sodium carbonate, which was prepared in Practical activity 2.4.

PURPOSE

To use the standard solution prepared in practical activity 2.4 to find the percentage, by mass, of hydrogen chloride (present as hydrochloric acid) in brick cleaner using an acid–base titration.

As directed by your teacher, complete the pre-lab safety information by referring to the hazard labels on the reagent bottles or safety data sheets (SDS) or your teacher's risk assessment for the activity.

PRE-LAB SAFETY INFORMATION

Material used	Hazard	Control
brick cleaner (concentrated hydrochloric acid)		
0.1 mol L^{-1} sodium carbonate solution		
methyl orange		

Please indicate that you have understood the information in the safety table.

Name (print): ______________________________

I understand the safety information (signature): ______________________________

MATERIALS

- 5 mL brick cleaner or concentrated hydrochloric acid, HCl
- 100 mL of approx. 0.1 mol L^{-1} standard sodium carbonate solution, Na_2CO_3 (The standard solution in practical activity 2.4 can be used.)
- 250 mL de-ionised water
- methyl orange indicator
- 250 mL volumetric flask
- 4 × 100 mL conical flasks
- small funnel
- 10 mL measuring cylinder
- 20 mL pipette
- pipette filler
- dropping pipette
- burette
- retort stand, bosshead and clamp
- white tile
- electronic balance
- tape and markers
- safety glasses

PROCEDURE

1 Weigh a clean, dry 250 mL volumetric flask and record its mass in Table 1.

2 Note the HCl content of the brick cleaner as specified by the manufacturer. Use a 10 mL measuring cylinder to pour about 5 mL of brick cleaner into the volumetric flask, avoiding spillages. Stopper the flask immediately.

3 Reweigh the flask plus contents and record the mass in Table 1.

4 Add de-ionised water until the flask is about half full. Stopper the flask and turn it upside down carefully several times to mix the solution thoroughly.

5 Add more water to the flask until the meniscus is level with the calibration line. Stopper and mix the solution thoroughly. Label the volumetric flask using the tape and a marker.

6 Use a 20 mL pipette to place 20.00 mL aliquots of standard, Na_2CO_3 solution into four 100 mL conical flasks. Add 2–3 drops of methyl orange indicator to each conical flask. Set one flask aside as a control in colour matching.

7 Fill a burette with the diluted solution of brick cleaner and record the initial level of the solution in the burette to two decimal places in Table 2.

8 Titrate the, Na_2CO_3 solution with the dilute solution of brick cleaner until the moment when the indicator just shows a permanent colour change from yellow to orange. Record the final burette reading in Table 2. Calculate the titre volume.

9 Repeat the titration, until you obtain three titres that are concordant (i.e. the smallest is no more than 0.1 mL less than the largest).

ISBN 978 1 4886 1933 5

RESULTS

TABLE 1 Brand, mass and concentration results

Brand of concrete cleaner	
HCl content specified by manufacturer	
Mass of 250 mL volumetric flask (g)	
Mass of 250 mL volumetric flask and brick cleaner (g)	
Concentration of standard Na_2CO_3 solution ($mol\ L^{-1}$)	

TABLE 2 Titration results

Titration	Initial burette reading (mL)	Final burette reading (mL)	Titre (mL)
1			
2			
3			
4			

PROCESSING DATA

1 Determine the mass of your sample of brick cleaner.

2 Find the average volume of your concordant titres.

3 Calculate the amount of sodium carbonate, in mol, present in each conical flask.

4 Write an equation for the reaction that occurs between hydrochloric acid and sodium carbonate solution.

5 Calculate the amount of hydrochloric acid, in mol, present in the average titre.

6 Calculate the amount of hydrochloric acid, in mol, present in the volumetric flask.

PRACTICAL ACTIVITY 2.5

7 Find the percentage, by mass, of hydrochloric acid in the brick cleaner.

ANALYSIS OF RESULTS

8 a How does your answer to question **7** compare with the manufacturer's claim about the composition of the brick cleaner? Suggest a reason for any differences.

b Compare your result with those of other members in your class. Explain why differences arise and how a more accurate result could be obtained.

DISCUSSION

9 Explain how you would safely clean up a spill of about 20 mL of brick cleaner on the floor.

CONCLUSION

RATING MY LEARNING	My understanding improved	Not confident ◄—► Very confident ○ ○ ○ ○ ○	I answered questions without help	Not confident ◄—► Very confident ○ ○ ○ ○ ○	I corrected my errors without help	Not confident ◄—► Very confident ○ ○ ○ ○ ○

ISBN 978 1 4886 1933 5

PRACTICAL ACTIVITY 2.6

Investigating the gas laws

Suggested duration: 80 minutes

INTRODUCTION

The ideal gas law links the gas properties of volume, pressure, amount and temperature. If two of these quantities are held constant, then the relationship between the remaining two quantities can be investigated. Practical investigations of this nature lead to Boyle's, Charles', Avogadro's and Gay-Lussac's laws.

MATERIALS

- 3 balloons

Part A—Volume-amount relationship (qualitative)

PURPOSE

To demonstrate the dependence of volume on the amount of a gas in a sample.

PROCEDURE

1 Blow a single breath into one of the balloons. Tie the neck of the balloon.

2 Blow two breaths into another balloon. Make sure no gas escapes between the breaths you take. Tie the neck of the balloon.

3 Blow three breaths into another balloon. Make sure no gas escapes between the breaths you take. Tie the neck of the balloon.

4 Think of a way to measure an aspect of the size of the three balloons. Record your observations of the relative sizes of the balloons in the results section.

RESULTS

ANALYSIS OF RESULTS

1 Which two quantities were kept constant in this experiment? How was this controlled?

DISCUSSION

2 How does the volume of a gas change as the amount of gas changes?

3 Which established gas law did this activity investigate?

Part B—Pressure-temperature relationship (qualitative) (teacher demonstration)

PURPOSE

To demonstrate the dependence of the vapour pressure of a liquid on temperature.

PRE-LAB SAFETY INFORMATION

Material used	Hazard	Control
steam and hot can	burns from steam and the hot can	Wear oven mitts when screwing the cap on the can.

Please indicate that you have understood the information in the safety table.

Name (print): ____________________

I understand the safety information (signature): ____________________

MATERIALS

- 4 L metal can with a tightly fitting screw cap
- Bunsen burner
- bench mat
- tripod
- 250 mL measuring cylinder
- oven mitts
- safety glasses

PROCEDURE

1 Place about 200 mL water in the can. With the cap off the can, boil the water using the Bunsen burner.

2 Allow steam to escape from the can for several minutes, then turn off the burner. Wearing oven mitts, immediately screw the lid on the can. Place the can on the bench.

3 As the can cools you will notice that it crushes. The process will be accelerated by running cold water over the outside of the can.

THEORY

When the water boils, a substantial proportion of the gas in the can is water vapour. Once the can is sealed and allowed to cool, this vapour condenses and the pressure in the can is reduced. Atmospheric pressure on the outside of the can then exceeds the internal gas pressure, causing the can to collapse.

ANALYSIS OF RESULTS

1 What was the purpose of boiling the water in the can?

____

2 Why was the can crushed as it cooled?

____

____

3 In which direction(s) does atmospheric pressure act on the can?

____

4 How does the vapour pressure of water change as the temperature increases?

____

5 Which established gas law did you investigate?

____

 ISBN 978 1 4886 1933 5

Part C—Volume-temperature relationship (qualitative)

PURPOSE

To examine qualitatively the relationship between the volume and temperature of a fixed amount of gas at constant pressure.

MATERIALS

- 3 balloons
- conical flask
- container of hot (not boiling) water
- container of ice water

PROCEDURE

1 Stretch the neck of the balloon over the opening of a conical flask.

2 Place the conical flask in a container of hot water and observe what happens to the balloon.

3 Place the conical flask in a container of ice water and observe what happens.

ANALYSIS OF RESULTS

1 How does the volume of air in the balloon change when the temperature is:

a increased?

____

b decreased?

____

2 What assumptions can you make about the amount of gas used in this investigation and its pressure?

____

____

3 Which established gas law did you investigate?

____

Part D—Volume-pressure relationships (quantitative)

PURPOSE

To develop a quantitative relationship between gas pressure and volume for a fixed amount of gas at constant temperature.

MATERIALS

- retort stand, clamp and bosshead
- 50 mL disposable plastic syringe
- rubber stopper with holes bored part-way through it
- 5 or 6 weights of equal mass (e.g. 500 g weights or half bricks weighing approx. 2 kg)
- safety glasses

PRE-LAB SAFETY INFORMATION

Material used	Hazard	Control
heavy weights	injury or damage due to falling weights	Conduct investigation on a tray and/or the floor (if clean).

Please indicate that you have understood the information in the safety table.

Name (print): _______________

I understand the safety information (signature): _______________

ISBN 978 1 4886 1933 5

PRACTICAL ACTIVITY 2.6

PROCEDURE

Record all of your procedure results in Table 1.

1 Clamp the syringe on to the retort stand with the nozzle downwards and the plunger set at full capacity. Fit the rubber stopper to the nozzle and record the volume of air in the syringe at full capacity.

2 Place a 500 g weight or half brick on top of the plunger and record the volume of air in the syringe. Check your result by repeating this step twice. Record each reading and calculate the average volume.

3 Place a second weight or half brick on top of the first and record the volume. Check and average your results as in step 2.

4 Repeat this procedure until a total of five or six weights have been used.

5 Calculate and record the reciprocal of each average volume.

RESULTS

TABLE 1 Volume results

Weight (kg)	**Volume (mL)**				**Reciprocal of volume $\left(\frac{1}{V}\right)$**
	Trial 1	**Trial 2**	**Trial 3**	**Average volume (mL)**	

PROCESSING DATA

1 Produce a graph by plotting average volume on the *y*-axis against pressure (represented by the mass of weights or bricks) on the *x*-axis. This is graph A. Join the points with a line of best fit.

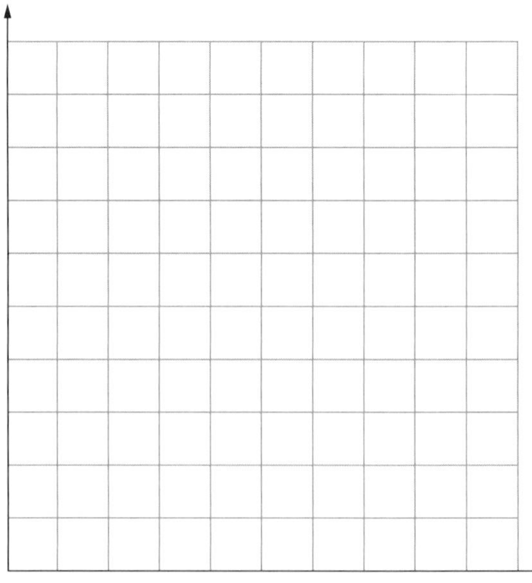

GRAPH A Average volume versus pressure

 ISBN 978 1 4886 1933 5

2 Plot the reciprocal of volume on the *y*-axis against pressure on the *x*-axis. This is graph B. Join the points with a line of best fit. To allow for the pressure that is caused by the atmosphere and by the weight of the plunger, raise the *x*-axis of graph B until the line passes through the origin.

GRAPH B Reciprocal of average volume versus pressure

ANALYSIS OF RESULTS

3 Use your observations to write a description of how the volume of a fixed amount of air at a fixed temperature varies with pressure.

4 What type of mathematical relationship between the volume of a gas and its pressure does graph B suggest?

5 Explain the relationship you have identified in questions **3** and **4**.

DISCUSSION

6 Which established gas law did you investigate in this experiment? Describe this law.

CONCLUSION

RATING MY LEARNING	My understanding improved	Not confident ◄──► Very confident ○ ○ ○ ○ ○	I answered questions without help	Not confident ◄──► Very confident ○ ○ ○ ○ ○	I corrected my errors without help	Not confident ◄──► Very confident ○ ○ ○ ○ ○

ISBN 978 1 4886 1933 5

DEPTH STUDY 2.1

How big is a mole? A primary- and secondary-sourced investigation

Suggested duration: 3.5 hours (including preparation of visual presentation)

INTRODUCTION

This depth study requires you to conduct an investigation into a question you develop about the size of a mole of a common object. You will research and process data and information. You will communicate your ideas in a digital, visual presentation.

The questions below are a guide to how you might proceed in the development and completion of this depth study. Your teacher may provide you with additional guidelines for the study and/or the presentation. The toolkit also provides advice on conducting a secondary-sourced investigation. GO TO ➤ page xviii

PURPOSE

This task requires you to use primary and/or secondary sources to investigate and answer two questions about the size of a mole of a common substance.

- One of the questions must be chosen from the list below.
- The second question must be developed yourself.

QUESTION 1

Choose one question from the following list. Highlight your chosen question.

- How many times to Jupiter and back would a tower containing a mole of ten-cent pieces reach?
- How many times the volume of the Earth would be occupied by a mole of tennis balls?
- How many times around the Earth's equator would a mole of toothpicks stretch if laid end to end?
- To what depth would a mole of square Lego bricks cover the Earth?

QUESTIONING AND PREDICTING

1 What data will you need to research to answer this question?

2 What calculations will you need to perform to process the data and answer this question?

3 Predict an answer to your question.

CONDUCTING YOUR INVESTIGATION

4 Record your researched data from primary or secondary sources in the following table. Record the details of the reference in the style advised by your teacher.

Relevant data	Source

5 The final product of this investigation will be a poster communicating your findings. A poster should always be visually appealing. Collect or draw some images that will be relevant to your questions and will help to communicate your answer to an audience.

Description of image	Source

 ISBN 978 1 4886 1933 5

QUESTION 2

You must develop your own question for question **2**. Choose a familiar object that interests you. You may be able to collect primary data about the size or dimensions of the object yourself.

Record your ideas in the space as you develop your question. Be sure to have the question approved by your teacher before you continue.

QUESTIONING AND PREDICTING

1 What data will you need to research to answer this question?

2 What calculations will you need to perform to process the data and answer this question?

3 Predict an answer to your question.

CONDUCTING YOUR INVESTIGATION

4 Record your researched data from primary or secondary sources in the following table. Record the details of the reference in the style advised by your teacher.

Relevant data	Source

5 The final product of this investigation will be a poster communicating your findings. A poster should always be visually appealing. Collect or draw some images that will be relevant to your questions and will help to communicate your answer to an audience.

Description of image	Source

PROCESSING DATA AND INFORMATION

Use the space to carry out calculations necessary to answer your questions.

COMMUNICATING

You are required to produce a digital presentation communicating your questions, data analysis and results to other students. Your presentation should include:

- statements of the problems or questions
- a list of the necessary information you needed to solve the problem, with an explanation of each
- calculations performed to answer the problem
- a bibliography.

Your presentation should be attractive in design and layout. You could use PowerPoint or a similar presentation program.

 ISBN 978 1 4886 1933 5

MODULE 2 • REVIEW QUESTIONS

Multiple choice

1 Which of the following samples contains the most atoms?

A 10.0 g of NO_2

B 10.0 g of CO_2

C 10.0 g of SO_3

D 10.0 g of CO

2 What is the mass of a single sulfur atom?

A $\frac{6.022 \times 10^{23}}{32.07}$

B $6.022 \times 10^{23} \times 32.07$

C $\frac{32.07}{6.022 \times 10^{23}}$

D $\frac{1}{6.022 \times 10^{23}}$

3 Which of the following compounds contains the greatest percentage by mass of sulfur?

A H_2SO_4

B SO_3

C $MgSO_4$

D $H_2S_2O_7$

4 What is the volume of 20.0 g of propane gas, C_3H_8, at 80.0 kPa and 5°C?

A 0.236 L

B 13.1 L

C 63.5 L

D 578 L

5 Which of the following solutions of sodium hydroxide, NaOH, is the most concentrated?

A 1.0 mol L^{-1}

B 30 g L^{-1}

C 500 ppm

D 2000 mg L^{-1}

6 Which of the following gases will occupy the biggest volume at SLC (standard laboratory conditions)?

A 44.0 g CO_2

B 32.0 g O_2

C 58.0 g C_4H_{10}

D They all have the same volume.

Short answer

7 Calculate the mass, in kg, of carbon dioxide gas produced from the combustion of 1.00 kg of butane gas according to the equation:

$2C_4H_{10}(g) + 13O_2(g) \rightarrow 8CO_2(g) + 10H_2O(g)$

8 Calculate the volume, in mL, of 2.00 mol L^{-1} nitric acid required to completely react with 3.45 g of magnesium metal according to the equation:

$Mg(s) + 2HNO_3(aq) \rightarrow Mg(NO_3)_2(aq) + H_2(g)$

Extended response

9 A student wishes to prepare 500 mL of a 0.100 mol L^{-1} standard solution of sodium carbonate, Na_2CO_3.

a Calculate the mass of sodium carbonate required.

b Write a method for the student to follow.

10 A student investigating the gas laws obtains the following data for an investigation of the relationship between volume and temperature.

Temperature (°C)	Volume (L)
0	2.20
20	2.43
40	2.58
60	2.76
80	2.92

a What properties of the gas would the student need to keep constant if her investigation is to be valid?

b Graph the student's results on the grid.

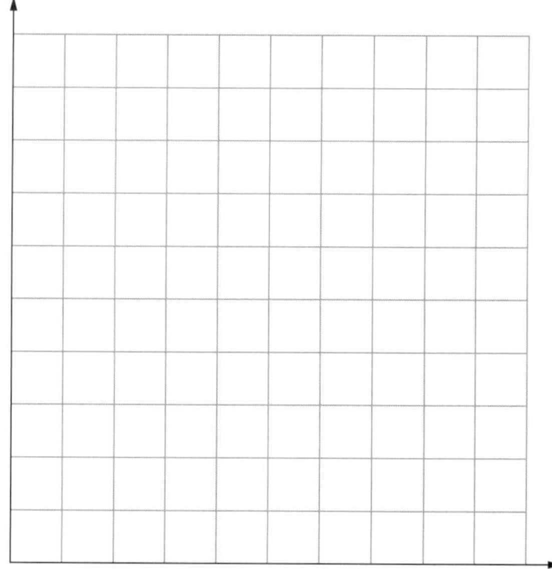

c What will be the volume of the gas at 35°C?

d **i** Which law has this student been investigating?

ii What is the mathematical expression of this law?

 ISBN 978 1 4886 1933 5

MODULE 3

Reactive chemistry

Outcomes

By the end of this module you will be able to:

- design and evaluate investigations in order to obtain primary and secondary data and information (CH11-2)
- conduct investigations to collect valid and reliable primary and secondary data and information (CH11-3)
- select and process appropriate qualitative and quantitative data and information using a range of appropriate media (CH11-4)
- explore the many different types of chemical reactions, in particular the reactivity of metals, and the factors that affect the rate of chemical reactions (CH11-10)

Content

CHEMICAL REACTIONS

INQUIRY QUESTION **What are the products of a chemical reaction?**

By the end of this module you will be able to:

- investigate a variety of reactions to identify possible indicators of a chemical change
- use modelling to demonstrate
  - the rearrangement of atoms to form new substances ICT
  - the conservation of atoms in a chemical reaction (ACSCH042, ACSCH080) ICT N
- conduct investigations to predict and identify the products of a range of reactions, for example:
  - synthesis
  - decomposition
  - combustion
  - precipitation
  - acid/base reactions
  - acid/carbonate reactions (ACSCH042, ACSCH080)
- investigate the chemical processes that occur when Aboriginal and Torres Strait Islander Peoples detoxify poisonous food items AHC
- construct balanced equations to represent chemical reactions ICT N

PREDICTING REACTIONS OF METALS

INQUIRY QUESTION **How is the reactivity of various metals predicted?**

By the end of this module you will be able to:

- conduct practical investigations to compare the reactivity of a variety of metals in:
  - water
  - dilute acid (ACSCH032, ACSCH037)
  - oxygen
  - other metal ions in solution

Module 3 • Reactive chemistry

- construct a metal activity series using the data obtained from practical investigations and compare this series with that obtained from standard secondary-sourced information (ACSCH103)
- analyse patterns in metal activity on the periodic table and explain why they correlate with, for example:
  - ionisation energy (ACSCH045)
  - atomic radius (ACSCH007)
  - electronegativity (ACSCH057)
- apply the definitions of oxidation and reduction in terms of electron transfer and oxidation numbers to a range of reduction and oxidation (redox) reactions
- conduct investigations to measure and compare the reduction potential of galvanic half-cells ICT
- construct relevant half-equations and balanced overall equations to represent a range of redox reactions
- predict the reaction of metals in solutions using the table of standard reduction potentials
- predict the spontaneity of redox reactions using the value of cell potentials (ACSCH079, ACSCH080)

RATES OF REACTIONS

INQUIRY QUESTION **What affects the rate of a chemical reaction?**

By the end of this module you will be able to:

- conduct a practical investigation, using appropriate tools (including digital technologies), to collect data, analyse and report on how the rate of a chemical reaction can be affected by a range of factors, including but not limited to: ICT N
  - temperature
  - surface area of reactant(s)
  - concentration of reactant(s)
- investigate the role of activation energy, collisions and molecular orientation in collision theory
- explain a change in reaction rate using collision theory (ACSCH003, ACSCH046) CCT

Key knowledge

Chemical reactions

CHEMICAL CHANGE

During a chemical reaction or chemical change, atoms are rearranged to make new substances. Chemical bonds are broken, new chemical bonds are made and energy transformations occur.

Possible indicators of a chemical change include:

- temperature increase
- temperature decrease
- emission of light
- formation of a solid, sometimes seen as cloudiness
- colour change
- noticeable odour after the reaction has started
- formation of a gas, often seen as bubbling.

A chemical change always involves the formation of new types of substance; however, it does not involve the formation of new types of atom. Rather, the atoms present in the reactants are rearranged to form the new products. The number and type of atom present in the reactant are conserved during a chemical change.

DIFFERENT TYPES OF CHEMICAL REACTIONS

Hundreds of different chemical reactions occur in our bodies, in the environment around us and in many areas of industry. Although chemical reactions are different, they can be grouped according to similarities in the way they proceed and the type of product formed. Different types of chemical reactions covered in this course include:

- synthesis
- decomposition
- combustion
- precipitation
- acid–base
- acid–carbonate.

Synthesis reactions

In a **synthesis reaction**, two or more simple reactants combine to form a more complex single product. The reactants may be elements or compounds; however, the product is always a compound.

For example:

$$2Mg(s) + O_2(g) \rightarrow 2MgO(s)$$
$$CaO(s) + CO_2(g) \rightarrow CaCO_3(s)$$

It is possible to have a second product in a synthesis reaction; however, this will be a simple substance alongside the more complex product.

For example:

$$6CO_2(g) + 6H_2O(l) \rightarrow C_6H_{12}O_6(aq) + 6O_2(g)$$

There are many different types of synthesis reactions (Table 3.1).

TABLE 3.1 Examples of synthesis reactions

Type of reactant	Type of product	Example
two elements	single compound that contains only two elements	$2Fe(s) + O_2(g) \rightarrow 2FeO(s)$
metallic oxide and carbon dioxide	single carbonate	$Na_2O(s) + CO_2(g) \rightarrow Na_2CO_3(s)$
metallic oxide and water	base	$Na_2O(s) + H_2O(l) \rightarrow 2NaOH(aq)$
non-metallic oxide and water	acid	$N_2O_5(g) + H_2O(l) \rightarrow 2HNO_3(aq)$

Decomposition reactions

In a **decomposition reaction**, a single compound is broken down into two or more products. The products can be elements or simple compounds.

For example:

$$2HgO(s) \rightarrow 2Hg(s) + O_2(g)$$
$$CaCO_3(s) \rightarrow CaO(s) + CO_2(g)$$

The different types of decomposition reaction include:

- thermal decomposition, in which the heating of a substance causes its breakdown into simpler substances
- electrical decomposition, also called electrolysis, in which electricity is used to cause decomposition
- photolysis, in which light causes decomposition to occur.

A decomposition reaction is often the reverse of a synthesis reaction. There are many different types of decomposition reactions (Table 3.2).

TABLE 3.2 Examples of decomposition reactions

Type of reactant	Type of product	Example
ionic compound that contains only two elements	two elements	$MgCl_2(s) \rightarrow Mg(s) + Cl_2(g)$
carbonate	metal oxide and carbon dioxide gas	$MgCO_3(s) \rightarrow MgO(s) + CO_2(g)$
acid	non-metal oxide and water	$H_2SO_4(l) \rightarrow SO_3(g) + H_2O(l)$
base	metal oxide and water	$2NaOH(aq) \rightarrow Na_2O(aq) + H_2O(l)$

Combustion reactions

Combustion is a reaction between a fuel and oxygen gas that releases energy. Most fuels are a compound containing at least carbon and hydrogen. The products of the complete combustion of a carbon- and hydrogen-based fuel are carbon dioxide and water.

For example:

$$\underset{\text{fuel}}{CH_4(g)} + 2O_2(g) \rightarrow CO_2(g) + 2H_2O(g)$$

Combustion also requires a small amount of energy to be provided to start the reaction, for example a match to light a fire.

If a combustion reaction takes place in an environment in which oxygen is limited, incomplete combustion can occur. The products of incomplete combustion can include carbon monoxide and carbon as well as water.

For example:

$$CH_4(g) + O_2(g) \rightarrow C(s) + 2H_2O(g)$$

Combustion reactions release energy in the form of heat and light.

Precipitation reactions

In a **precipitation reaction** an insoluble compound is formed when two clear solutions are mixed together.

Solubility of ionic substances

Not all ionic substances dissolve well in water, although the ability to dissolve tends to increase with increasing water temperature. A **solubility table** (Table 3.3) summarises the solubility in water of many common ionic compounds. It is useful to remember that:

- all compounds containing a nitrate ion are soluble
- all compounds of group 1 elements are soluble.

When two solutions are mixed an insoluble substance called a **precipitate** sometimes forms (Figure 3.1). The balanced chemical equation for a precipitation reaction has the general form:

ionic compound(aq) + ionic compound(aq) → precipitate(s) + ionic compound(aq)

For example:

$$Pb(NO_3)_2(aq) + 2KI(aq) \rightarrow PbI_2(s) + 2KNO_3(aq)$$

In this case, PbI_2 is the precipitate. K^+ and NO_3^- ions remain dissolved in solution and are termed **spectator ions**.

An **ionic equation** omits spectator ions to give a more accurate picture of the reaction taking place.

For example:

$$Pb^{2+}(aq) + 2I^-(aq) \rightarrow PbI_2(s)$$

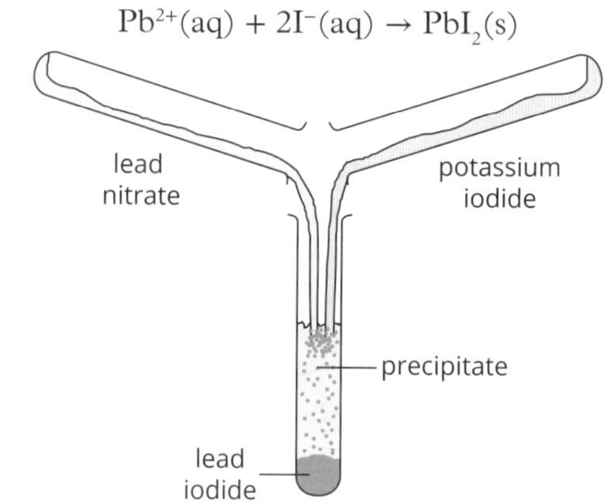

FIGURE 3.1 Mixing solutions of lead nitrate and potassium iodide forms a precipitate of lead iodide.

The product that occurs during a precipitation reaction can be observed as cloudiness, a milky appearance or a distinct solid being produced as the reaction proceeds. There may also be a change in colour and/or temperature.

TABLE 3.3 Solubility of some ionic compounds in water

Soluble in water (>0.1 mol dissolves per L at 25°C)	Exceptions: insoluble (<0.01 mol dissolves per L at 25°C)	Exceptions: slightly soluble (0.01–0.1 mol dissolves per L at 25°C)
all nitrates, NO_3^-	no exceptions	no exceptions
all ammonium, NH_4^+, salts	no exceptions	no exceptions
all sodium, Na^+, and potassium, K^+, salts	no exceptions	no exceptions
all acetates (ethanoates), CH_3COO^-	no exceptions	no exceptions
most sulfates, SO_4^{2-}	$BaSO_4$, $PbSO_4$, $SrSO_4$	$CaSO_4$, Ag_2SO_4
most chlorides, Cl^-, bromides, Br^-, and iodides, I^-	AgCl, AgBr, AgI, PbI_2	$PbCl_2$, $PbBr_2$
Insoluble in water	**Exceptions: soluble**	**Exceptions: slightly soluble**
most hydroxides, OH^-	NaOH, KOH, $Ba(OH)_2$, NH_4OH	$Ca(OH)_2$, $Sr(OH)_2$
most carbonates, CO_3^{2-}	Na_2CO_3, K_2CO_3, $(NH_4)_2CO_3$	no exceptions
most phosphates, PO_4^{3-}	Na_2PO_4, K_2PO_4, $(NH_4)_2PO_4$	no exceptions
most sulfides, S^{2-}	Na_2S, K_2S, $(NH_4)_2S$	no exceptions

 ISBN 978 1 4886 1933 5

REACTIONS OF ACIDS AND BASES

Acids and bases are groups of solutions that have characteristic properties (Table 3.4).

TABLE 3.4 Properties of acids and bases

Properties of acids	Properties of bases
taste sour	taste bitter
turn litmus paper red	turn litmus paper blue
can be strong or weak	can be strong or weak
can be dilute or concentrated	can be dilute or concentrated
react with bases to produce salt and water	react with acids to produce salt and water
can be corrosive	can be corrosive

Acids and reactive metals

When dilute acids are added to main-group metals and some transition metals, bubbles of hydrogen gas are released and a salt is formed.

For example:

$$2HCl(aq) + Mg(s) \rightarrow MgCl_2(aq) + H_2(g)$$

The ionic equation is:

$$2H^+(aq) + Mg(s) \rightarrow Mg^{2+}(aq) + H_2(g)$$

Acids and metal carbonates

A carbonate is an ionic compound that contains a carbonate ion, CO_3^{2-}. Examples are sodium carbonate, Na_2CO_3, magnesium carbonate, $MgCO_3$, and copper(II) carbonate, $CuCO_3$.

When a dilute acid reacts with a metal carbonate, the products are a salt, carbon dioxide and water.

For example:

$$\underset{\text{acid}}{2HCl(aq)} + \underset{\text{carbonate}}{Na_2CO_3(aq)} \rightarrow \underset{\text{salt}}{2NaCl(aq)} + \underset{\text{water}}{H_2O(l)} + \underset{\text{carbon dioxide}}{CO_2(g)}$$

The ionic equation is:

$$2H^+(aq) + CO_3^{2-}(aq) \rightarrow H_2O(l) + CO_2(g)$$

The change that occurs during an acid–carbonate reaction can be observed by the bubbling due to production of CO_2. The identity of the gas as carbon dioxide can be tested by bubbling the gas through lime water. Only carbon dioxide gas will make the lime water cloudy. There may also be an observable change in temperature.

REMOVING TOXINS FROM FOODS

Indigenous Australians have employed chemical processes for thousands of years to remove toxins as part of their food preparation. Following are two examples.

- The fruit of the cycad is highly toxic and carcinogenic. Indigenous Australians would soak the fruit in water (leaching) and ferment it to remove the toxins.
- Toxic black beans were heated, scraped and leached to make them edible.

Predicting reactions of metals

REACTIONS OF METALS

Metals can be reacted with oxygen, water, dilute acid and/or other metal ions in solution in order to experimentally determine their relative reactivities.

- When a metal reacts with oxygen it forms an oxide. Naturally occurring metals in the Earth's crust often exist as oxides. Very unreactive metals, such as gold and platinum, exist in the Earth's crust in their pure form.
- When a metal reacts with water it produces hydrogen gas. The presence and vigour of the bubbling are an indication of the metal's reactivity.
- Metals that do not react with water may react with a dilute acid such as HCl. Hydrogen gas is a product, so again the presence and vigour of bubbling are an indication of the metal's reactivity.
- Metals can also be reacted with other metal ions in solution in order to determine their relative reactivity. A more reactive metal will displace the ion of a less reactive metal from solution. For example, zinc is more reactive than copper so zinc will react with copper ions in solution. However, copper will not react with zinc ions in solution.

The results from a number of experiments can be collated to summarise the reactivity of metals (Table 3.5).

TABLE 3.5 Summary of reactivity of metals with oxygen, water, dilute acid and solutions of metal ions

<table>
<tr><th rowspan="2">Metal</th><th colspan="4">Reaction with</th></tr>
<tr><th>oxygen</th><th>water</th><th>dilute acid</th><th>solution of metal ions</th></tr>
<tr><td>K</td><td rowspan="15">react with oxygen to form a metal oxide</td><td rowspan="4">displace hydrogen from cold water</td><td rowspan="4">not tested</td><td rowspan="4">not tested</td></tr>
<tr><td>Na</td></tr>
<tr><td>Li</td></tr>
<tr><td>Ca</td></tr>
<tr><td>Mg</td><td rowspan="6">displace hydrogen from steam</td><td rowspan="10">displace hydrogen from dilute acids</td><td rowspan="12">displace the ion of a metal lower in the series from a solution of its salt</td></tr>
<tr><td>Al</td></tr>
<tr><td>Mn</td></tr>
<tr><td>Zn</td></tr>
<tr><td>Cr</td></tr>
<tr><td>Fe</td></tr>
<tr><td>Co</td><td rowspan="6">do not displace hydrogen from either cold water or steam</td></tr>
<tr><td>Ni</td></tr>
<tr><td>Sn</td></tr>
<tr><td>Pb</td></tr>
<tr><td>Cu</td><td rowspan="2">do not displace hydrogen from dilute acids</td></tr>
<tr><td>Ag</td><td>does not react with oxygen</td></tr>
</table>

THE ACTIVITY SERIES OF METALS

An **activity series of metals** (Table 3.6) lists metals in order of their relative reactivity based on experimental data. Group 1 metals are the most reactive metals while transition metals are the least reactive. Metals at the top of the activity series lose electrons more readily than metals lower down the series and consequently are more reactive.

TABLE 3.6 Activity series of metals with half-equations

Order of activity of metal	Half-equation representing loss of electrons
decreasing metal activity ↓	$K(s) \rightarrow K^+(aq) + e^-$
	$Na(s) \rightarrow Na^+(aq) + e^-$
	$Li(s) \rightarrow Li^+(aq) + e^-$
	$Ca(s) \rightarrow Ca^{2+}(aq) + 2e^-$
	$Mg(s) \rightarrow Mg^{2+}(aq) + 2e^-$
	$Al(s) \rightarrow Al^{3+}(aq) + 3e^-$
	$Mn(s) \rightarrow Mn^{2+}(aq) + 2e^-$
	$Zn(s) \rightarrow Zn^{2+}(aq) + 2e^-$
	$Fe(s) \rightarrow Fe^{2+}(aq) + 2e^-$
	$Cr(s) \rightarrow Cr^{3+}(aq) + 3e^-$
	$Co(s) \rightarrow Co^{2+}(aq) + 2e^-$
	$Ni(s) \rightarrow Ni^{2+}(aq) + 2e^-$
	$Sn(s) \rightarrow Sn^{2+}(aq) + 2e^-$
	$Pb(s) \rightarrow Pb^{2+}(aq) + 2e^-$
	$Cu(s) \rightarrow Cu^{2+}(aq) + 2e^-$
	$Ag(s) \rightarrow Ag^+(aq) + e^-$
	$Au(s) \rightarrow Au^+(aq) + e^-$

A reaction will only occur between a metal and the ion of another metal if the ion is of a less reactive metal.

The activity series is based on experimental observations. It has limitations because factors such as temperature, surface area of solids and concentration of reactants will also affect how vigorously a reaction occurs.

Standard reduction potentials

A table of standard reduction potentials (Table 3.7 and page 105) ranks both metals and non-metals in order of reactivity. It also includes a value for electrode potential (in volts), which has been measured under specific conditions of:

- temperature (25°C)
- concentration (1 mol L^{-1} solutions)
- pressure (100 kPa).

TABLE 3.7 Section of the table of standard reduction potentials containing different metals

Half-equation	Electrode potential (V)
$K^+(aq) + e^- \rightarrow K(s)$	−2.94
$Ca^{2+}(aq) + 2e^- \rightarrow Ca(s)$	−2.87
$Na^+(aq) + e^- \rightarrow Na(s)$	−2.71
$Mg^{2+}(aq) + 2e^- \rightarrow Mg(s)$	−2.36
$Al^{3+}(aq) + 3e^- \rightarrow Al(s)$	−1.68
$Zn^{2+}(aq) + 2e^- \rightarrow Zn(s)$	−0.76
$Fe^{2+}(aq) + 2e^- \rightarrow Fe(s)$	−0.44
$Sn^{2+}(aq) + 2e^- \rightarrow Sn(s)$	−0.14
$Pb^{2+}(aq) + 2e^- \rightarrow Pb(s)$	−0.13
$Cu^{2+}(aq) + 2e^- \rightarrow Cu(s)$	+0.34
$Ag^+(aq) + e^- \rightarrow Ag(s)$	+0.80

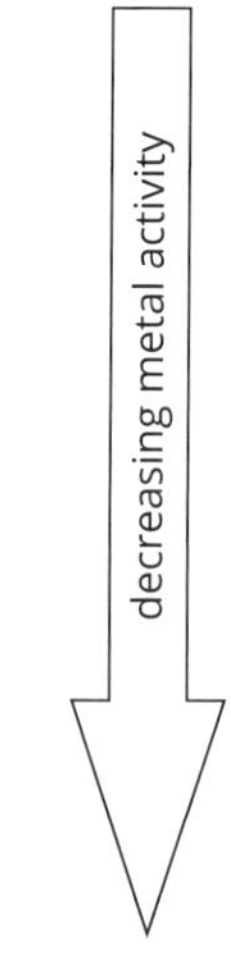

METAL ACTIVITY AND THE PERIODIC TABLE

In Module 1, you learnt about some of the trends in properties of elements in the periodic table. The reactivity of metals correlates with some of these trends including: GO TO ➤ page 8

- ionisation energy
- atomic radius
- electronegativity (Figure 3.2).

The trends correlate because each of these properties, including reactivity, is a reflection of how readily a metal atom will lose its valence electron(s):

- From left to right across a period, core charge is increasing, which means the valence electrons are held more tightly. Ionisation energy increases, electronegativity increases, atomic radius decreases and reactivity of metals decreases.
- From top to bottom down a group, the core charge stays the same, but the valence electrons are increasingly further from the nucleus as there are more occupied shells. This means the valence electrons are held less tightly. Ionisation energy decreases, electronegativity decreases, atomic radius increases and metals become more reactive.

ISBN 978 1 4886 1933 5

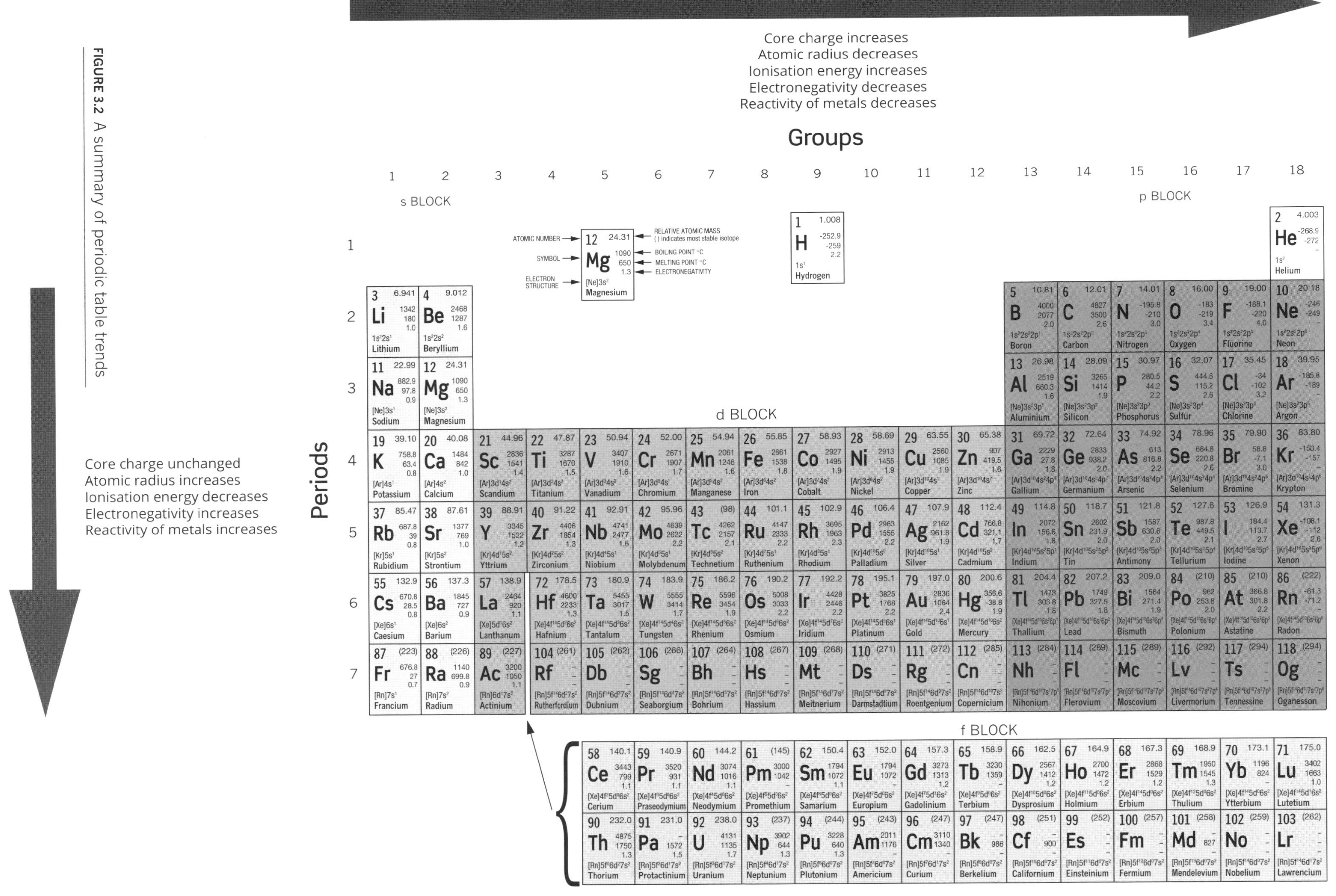

FIGURE 3.2 A summary of periodic table trends

ISBN 978 1 4886 1933 5

REDOX REACTIONS AND GALVANIC CELLS

Introducing redox reactions

Redox reactions (**red**uction–**ox**idation) involve the transfer of electrons.

Redox reactions occur in two parts:

- **oxidation**—loss of electrons
- **reduction**—gain of electrons.

Oxidation and reduction occur simultaneously. The electrons lost by the species undergoing oxidation are the electrons gained by the species undergoing reduction. The species undergoing oxidation is called the **reducing agent** or reductant. The species undergoing reduction is called the **oxidising agent** or oxidant (Figure 3.3).

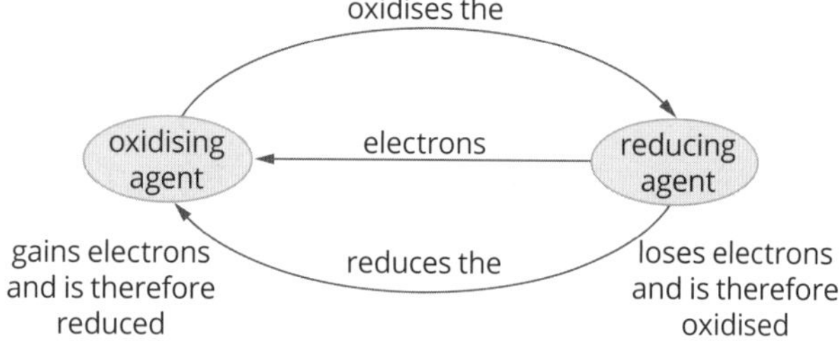

FIGURE 3.3 Relationship between oxidising and reducing agents

The oxidising agent and its product after it gains electrons, for example Cu^{2+} and Cu, are referred to as a conjugate redox pair. Redox reactions always involve two conjugate redox pairs.

Writing redox reactions

Half-equations are written to represent the separate processes of oxidation and reduction. Half-equations are balanced in terms of:

- elements—the number and type of atoms on the reactant side of the equation must equal the number and type on the product side
- charge—electrons are represented by the symbol e^- and are included to adjust the charge on either the reactant or product side as needed.

States are also included.

For example, in a reaction between copper metal and silver chloride, copper is oxidised to copper ions and silver ions are reduced.

Oxidation: $Cu(s) \rightarrow Cu^{2+}(aq) + 2e^-$ ↑ reducing agent	Electrons are lost, so appear as a product. They balance the charge on the copper ion. 2 electrons are lost for each copper atom oxidised.
Reduction: $Ag^+(aq) + e^- \rightarrow Ag(s)$ ↑ oxidising agent	An electron is gained, so it appears as a reactant. It balances the charge on the silver ion. 1 electron is gained to reduce each silver ion.

An overall equation is written as a combination of half-equations. The overall equation is balanced so that the number of electrons lost in the oxidation reaction is equal to the number of electrons gained in the reduction reaction. In the previous example, two silver ions will be reduced for every copper atom oxidised. Electrons do not appear in the overall equation:

$$Cu(s) + 2Ag^+(aq) \rightarrow Cu^{2+}(aq) + 2Ag(s)$$

More complex half-equations occurring in acidic solutions are balanced following the rules in Table 3.8.

TABLE 3.8 Rules for balancing complex redox half-equations in acidic solutions

Rule	Example: reduction of $Cr_2O_7^{2-}(aq)$ to $Cr^{3+}(aq)$
1 Balance all elements except hydrogen and oxygen.	$Cr_2O_7^{2-} \rightarrow 2Cr^{3+}$
2 Balance oxygen atoms using H_2O molecules.	$Cr_2O_7^{2-} \rightarrow 2Cr^{3+} + 7H_2O$
3 Balance hydrogen atoms using H^+ ions.	$Cr_2O_7^{2-} + 14H^+ \rightarrow 2Cr^{3+}(aq) + 7H_2O$
4 Balance charge using electrons, and add states.	$Cr_2O_7^{2-}(aq) + 14H^+(aq) + 6e^- \rightarrow 2Cr^{3+}(aq) + 7H_2O(l)$

Oxidation numbers

Oxidation is also defined as occurring when there is an increase in **oxidation number**. Reduction is occurring when there is a decrease in oxidation number.

Oxidation numbers:

- are used to decide if a reaction is a redox reaction
- have no physical meaning; they are defined to enable reactions involving molecules as well as those involving ions to be described as redox reactions
- are determined using the rules in Table 3.9.

TABLE 3.9 Rules for determining oxidation numbers

Species	Oxidation number	Examples
elements	0	Cl_2, Mg, C, O_2, H_2
ions	charge on the ion	$Na^+ = +1$; written as $\overset{+1}{Na^+}$ $Cl^- = -1$; written as $\overset{-1}{Cl^-}$
oxygen in compounds	defined as −2 in its compounds; H_2O_2 is an exception, where it is −1	O = −2 in H_2O, CO_2, Na_2O Written as $H_2\overset{-2}{O}$, $C\overset{-2}{O}_2$, $Na_2\overset{-2}{O}$
hydrogen in compounds	defined as +1 in compounds with non-metals; exception is −1 in compounds with metals	$\overset{+1}{H}Cl$, $\overset{+1}{H}_2S$, $C\overset{+1}{H}_4$ $Na\overset{-1}{H}$
molecular ions and molecules	The sum of the oxidation numbers equals the charge on the molecular ion (zero in the case of neutral molecules). The most electronegative element has the negative oxidation number.	For MnO_4^- oxygen is defined as −2. Because there are 4 oxygen atoms, in order to have an overall charge on the MnO_4^- ion of −1, Mn must have an oxidation number of +7. Let oxidation number of Mn be x: $x + 4(-2) = -1$ So $x = +7$

ISBN 978 1 4886 1933 5

GALVANIC CELLS

Galvanic cells are electrochemical cells that convert chemical energy to electrical energy. They involve **spontaneous redox reactions**. The two half-reactions occur in two separate **half-cells**. This prevents direct contact between the oxidising agent and reducing agent, so electrons can only be transferred by travelling through an external circuit, between the negative and positive electrodes. The flow of electrons provides electrical energy.

Galvanic cells have the same general features (Figure 3.4).

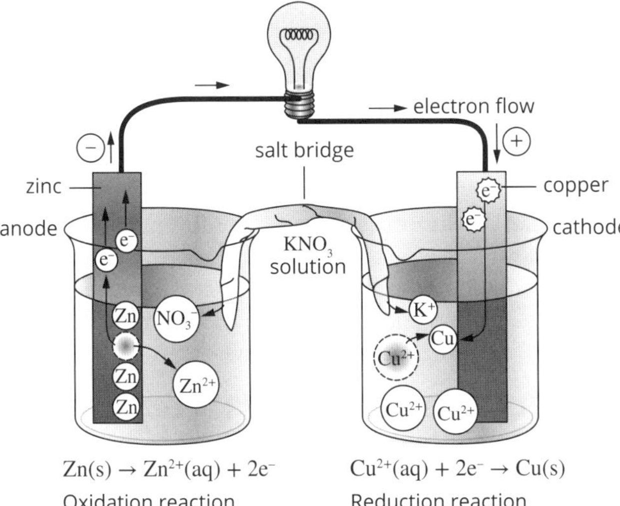

FIGURE 3.4 General features of a galvanic cell

In the cell in Figure 3.4, the zinc metal electrode is oxidised to zinc ions:

$$Zn(s) \rightarrow Zn^{2+}(aq) + 2e^-$$

The electrons produced move through the wire (the external circuit) towards the positive electrode, where they reduce copper ions:

$$Cu^{2+}(aq) + 2e^- \rightarrow Cu(s)$$

The number of electrons produced is equal to the number consumed at the other half-cell. The overall equation is the sum of the two half-equations:

$$Cu^{2+}(aq) + Zn(s) \rightarrow 2Cu(s) + Zn^{2+}(aq)$$

Zn is oxidised and acts as the reducing agent. Cu^{2+} is reduced and acts as the oxidising agent.

The **salt bridge** (often a piece of filter paper soaked in a solution of a soluble ionic compound) balances the charges formed and consumed in each half-cell. At the negative electrode, the positive Zn^{2+} ions being formed are balanced by negative ions moving from the salt bridge. At the positive electrode, the Cu^{2+} ions being consumed are balanced by positive ions from the salt bridge.

By definition, the electrode where oxidation occurs is the **anode** and the electrode where reduction occurs is the **cathode**. In galvanic cells, the anode is negative and the cathode is positive. (You can remember oxidation occurs at the anode because oxidation and anode start with vowels; both reduction and cathode start with consonants.)

Each half-cell contains a conjugate redox pair. In the cell in Figure 3.4 the pairs can be written as $Zn^{2+}(aq)/Zn(s)$ and $Cu^{2+}(aq)/Cu(s)$. In the salt bridge, cations move towards the cathode and anions move towards the anode.

TABLE OF STANDARD REDUCTION POTENTIALS

The **table of standard reduction potentials** (page 102) shows the relative strengths of oxidising agents and reducing agents. It is a more complex version of the activity series of metals. It contains non-metallic elements and compounds as well as values for the standard reduction potential of each half-cell. Half-reactions are still written as reduction reactions, with the strongest oxidising agent at the bottom of the table on the left-hand side and the strongest reducing agent at the top of the table on the right-hand side.

The **standard hydrogen half-cell** is used as a reference (Figure 3.5) and given an E° value of 0V. The half-equation for the cell is:

$$2H^+(aq) + 2e^- \rightarrow H_2(g)$$

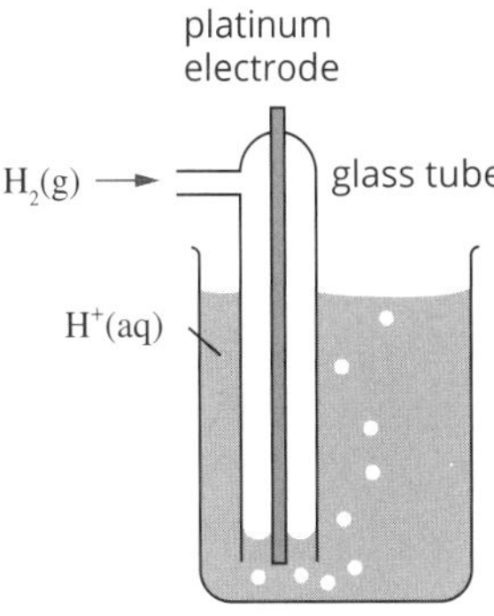

FIGURE 3.5 Standard hydrogen half-cell

The table of standard reduction potentials is constructed by measuring the voltages of different half-cells when they are connected to the standard hydrogen electrode. Equations for half-reactions that occur in half-cells that are positive with respect to the standard hydrogen electrode are below the $H^+(aq)/H_2(g)$ equation in the series and those that are negative are above it.

The cell potentials for any combination of two half-cells can be calculated using the E° values from the electrochemical series.

$E^\circ_{cell} = E^\circ$(of half-equation containing oxidising agent) – E°(of half-equation containing reducing agent)

Using the table of standard reduction potentials

A spontaneous reaction occurs when an oxidising agent comes in contact with a reducing agent that is higher in the table of standard reduction potentials. This means the E° of the oxidising agent is more positive (lower in the table) than the E° of the reducing agent. If there are several possible oxidising agents and reducing agents present, the pair that is most likely to react will be furthest apart in the table.

ISBN 978 1 4886 1933 5

Worked example 3.1: A cell is made from $Ag^+(aq)/Ag(s)$ and $Zn^{2+}(aq)/Zn(s)$ half-cells. Use the table of standard reduction potentials to predict the electrode reactions, write the overall cell equation and determine the maximum cell voltage under standard conditions.

GO TO ➤ page 102

Thinking	Working
Write the relevant half-equations.	$Zn^{2+} + 2e^- \rightarrow Zn$ $Ag^+ + e^- \rightarrow Ag$
Of the chemicals in the cell, the chemical lowest on the left of the table of standard reduction potentials (the strongest oxidising agent) will react with the chemical highest on the right of the table of standard reduction potentials (the strongest reducing agent). Write the anode reaction and the cathode reaction.	Ag^+ is the strongest oxidising agent. Zn is the strongest reducing agent. So Ag^+ and Zn will react. Anode reaction: $Zn \rightarrow Zn^{2+} + 2e^-$ Cathode reaction: $Ag^+ + e^- \rightarrow Ag$
Write the overall cell equation, including states.	$2Ag^+(aq) + Zn(s) \rightarrow 2Ag(s) + Zn^{2+}(aq)$
Calculate the cell potential.	$E°cell = E°$ at cathode − $E°$ at anode = +0.80 − (−0.76) = 1.56 V

Limitations of predictions using the table of standard reduction potentials

There are two significant limitations with the use of the table of standard reduction potentials:

- It applies to reactions under standard conditions. For non-standard conditions, the series cannot be used reliably to predict reactions.
- It predicts the likelihood of a reaction, but not the rate of reaction. A reaction may not be observed when chemicals are directly mixed together if the reaction occurs slowly.

Rates of reaction

COLLISION THEORY

Factors that affect the rate of reaction can be explained by **collision theory**. Collision theory describes that a chemical reaction will only occur when:

- reactant particles collide
- the collision has sufficient energy and the reactant particles are in the correct orientation to break the bonds in the reactants.

Not all collisions between reactant particles are successful, that is, result in a reaction. The likelihood of a reaction occurring when particles collide depends on the:

- energy of particles during collision, which must be equal to or greater than the activation energy
- molecular orientation or arrangement of the particles when they collide.

The **activation energy** is the minimum energy required for a reaction to occur. It is energy required to break the bonds in the reactant particles. A reaction with a lower activation energy will have a faster reaction rate than if it had a higher activation energy.

MEASURING REACTION RATE

The rate of a chemical reaction can be determined by measuring the change in the amount or concentration of the reactants or products with time. The unit of reaction rate is $mol\,L^{-1}\,s^{-1}$. This could include measurement of the rate of change of:

- volume of gas produced
- mass loss
- change in temperature
- change in colour
- change in pH
- change in electrical conductivity.

 ISBN 978 1 4886 1933 5

EFFECT OF SURFACE AREA, CONCENTRATION AND PRESSURE ON REACTION RATE

Reactions can only occur when reactant particles collide, so reaction rate depends on the frequency of collisions (number of collisions per second). Any change that causes collisions to occur more frequently will increase the rate of reaction.

Factors that affect the rate of a chemical reaction by altering the frequency of collisions include:

- surface area of reactant(s)
- concentration of reactant(s)
- pressure of gaseous reactant(s).

Collision theory can be used to explain the changes in reaction rate that occur when one of these factors is altered (Table 3.10).

TABLE 3.10 Factors affecting the rate of reaction by changing frequency of collisions

Factor	Effect on rate	
increasing concentration or pressure	An increased concentration of reactant particles causes collisions to occur more frequently. This results in faster formation of products. For reactions involving gases, an increase in pressure as a result of a decrease in volume has a similar effect.	Low concentration Higher concentration
increasing surface area of solids	Increasing the surface area makes more particles available to collide. This increases the frequency of the reactant particles colliding, and the rate of the formation of products increases.	Log of wood (1 piece) (smoking only) Many thin pieces of wood (flames and smoke)

EFFECT OF TEMPERATURE ON REACTION RATE

At any particular temperature, the particles in a substance have a range of kinetic energies. When the temperature of the substance increases, the average kinetic energy of the particles increases (Table 3.11).

The increased kinetic energy of the particles means that collisions occurring at higher temperatures have greater energy than those at lower temperatures. More colliding particles will have energies that are greater than or equal to the activation energy and so the proportion of successful collisions also increases.

As the temperature of a reaction increases, the increased speed of particles causes them to collide more often, increasing the frequency of successful collisions and the rate of reaction. However, the greater proportion of successful collisions is the main reason why increasing temperature increases the rate.

TABLE 3.11 Factors affecting the rate of reaction by changing proportion of successful collisions

increasing temperature	At any specified temperature, particles have a range of velocities and kinetic energies. Increasing the temperature causes more reactant particles to have more energy and so collide more frequently. More importantly, a higher proportion will have kinetic energies greater than the activation energy (E_a). The result is faster product formation.	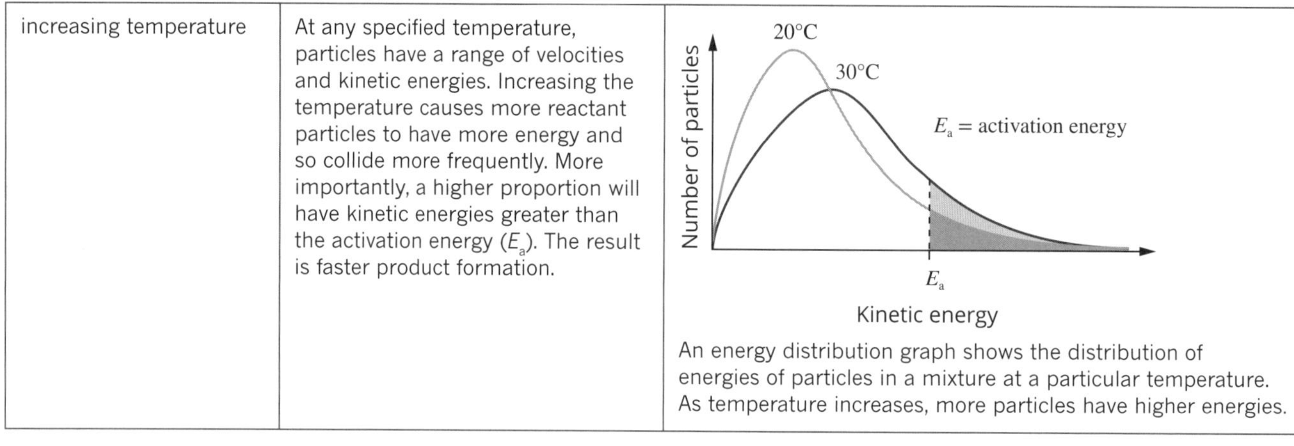 An energy distribution graph shows the distribution of energies of particles in a mixture at a particular temperature. As temperature increases, more particles have higher energies.

 ISBN 978 1 4886 1933 5

WORKSHEET 3.1

Knowledge review—identifying and naming types of substances, balancing chemical equations

1 A chemical reaction involves the rearrangement of particles in a substance to make up a new substance. Bonds must be broken and reformed. Identify the type(s) of bonding present in each of the substances in the table.

Substance	Type(s) of bonding
magnesium	
aluminium nitride	
carbon monoxide	
hydrogen chloride	
oxygen	
iron	
iron oxide	
carbon dioxide	
water (dihydrogen monoxide)	
diamond	
sodium hydroxide	

2 When a chemical reaction occurs, mass is conserved.

a Which subatomic particles make a significant contribution to the mass of an atom?

____

b In the following reaction, 5.0 g of magnesium reacts with oxygen to produce magnesium oxide:

$2Mg(s) + O_2(g) \rightarrow 2MgO(s)$

Calculate the mass of oxygen that will react with the magnesium.

____

____

____

c Without using stoichiometry, predict the mass of magnesium oxide formed. Give a reason for your prediction.

____

____

3 Balanced chemical equations demonstrate that the type and number of atoms in the reactants are also present in the products. Balance each of the following equations.

__$HCl(aq)$ + __$MgCO_3(aq)$ → __$MgCl_2(aq)$ + __$H_2O(l)$ + __$CO_2(g)$

__$C_4H_{10}(g)$ + __$O_2(g)$ → __$CO_2(g)$ + __$H_2O(g)$

__$Al(s)$ + __$O_2(g)$ → __$Al_2O_3(s)$

ISBN 978 1 4886 1933 5

WORKSHEET 3.2

Predicting products—synthesis, decomposition and combustion

A student conducts an investigation into the products of decomposition, synthesis and combustion reactions. The student's results are in the table. For each experiment, identify the type of reaction that has occurred as synthesis, decomposition or combustion, name the product(s) and write a balanced chemical equation for the reaction. In each case, justify your choice.

Investigation and observations	Type of reaction	Name(s) of product(s)	Balanced chemical equation	Justification of choice
The student observes the burning of propane gas, C_3H_8, in a barbecue. She observes an orange-blue flame and can feel heat.				
The student pours some hydrogen peroxide, H_2O_2, into a measuring flask. She adds some detergent and then a spoonful of catalyst. Vigorous bubbling is observed, producing large amounts of foam.				
The student reacts a solid sodium-containing compound with water that contains pH indicator. The pH change indicates a base has formed.				
The student leaves some magnesium on a bench for a long period of time. She observes the formation of grey-black solid on the surface of the magnesium.				
The student observes the reaction when methanol, CH_3OH, is used to heat water in a camping stove. She observes an orange-blue flame and can feel heat.				

RATING MY LEARNING	My understanding improved	Not confident ◄—► Very confident ○ ○ ○ ○ ○	I answered questions without help	Not confident ◄—► Very confident ○ ○ ○ ○ ○	I corrected my errors without help	Not confident ◄—► Very confident ○ ○ ○ ○ ○

ISBN 978 1 4886 1933 5

WORKSHEET 3.3

Solving solubility—predicting precipitation reactions

A solubility table summarises the solubility in water of many common ionic compounds.

Soluble ionic compounds	
Most compounds of the following ions are soluble	**Important exceptions**
nitrates (NO_3^-)	none
acetates (CH_3COO^-)	none
sodium, potassium (Na^+, K^+)	none
ammonium (NH_4^+)	none
chlorides (Cl^-)	$AgCl$, $HgCl_2$, $PbCl_2$—slightly soluble
bromides (Br^-)	$AgBr$, $HgBr_2$, $PbBr_2$—slightly soluble
iodides (I^-)	AgI, HgI_2, PbI_2
sulfates (SO_4^{2-})	$BaSO_4$, $PbSO_4$, $CaSO_4$—slightly soluble
Insoluble ionic compounds	
Most compounds of the following ions are insoluble	**Important exceptions**
sulfides (S^{2-})	Na_2S, K_2S, Li_2S, $(NH_4)_2S$
carbonates (CO_3^{2-})	Na_2CO_3, K_2CO_3, Li_2CO_3
phosphates (PO_4^{3-})	Na_3PO_4, K_3PO_4, Li_3PO_4
hydroxides (OH^-)	$NaOH$, KOH, $Ba(OH)_2$, $Ca(OH)_2$—slightly soluble

1 The following solutions are mixed together. Indicate whether or not you think a precipitate will form by placing a tick or cross in each box in the second column. For each predicted precipitate, write a full, balanced chemical equation in the third column.

Solutions mixed	Precipitate? (✓ or ✗)	Balanced chemical equation
potassium chloride + silver nitrate		
copper(II) nitrate + sodium hydroxide		
magnesium nitrate + sodium chloride		
lead nitrate + potassium chloride		
sodium sulfate + calcium nitrate		

2 Outline an experimental method a student could follow to obtain a dry sample of calcium carbonate from solutions of calcium chloride and sodium carbonate.

RATING MY LEARNING	My understanding improved	Not confident ←→ Very confident ○ ○ ○ ○ ○	I answered questions without help	Not confident ←→ Very confident ○ ○ ○ ○ ○	I corrected my errors without help	Not confident ←→ Very confident ○ ○ ○ ○ ○

ISBN 978 1 4886 1933 5

WORKSHEET 3.4

Reactions of acids—predicting products

Acids demonstrate similar chemical behaviour. They react with many types of substance, including bases and carbonates, in common ways.

1 Complete the table by predicting the specific products of the following reactions.

Reaction number	Reactants	Products
1	hydrochloric acid and a solution of sodium hydroxide	
2	nitric acid and aqueous potassium oxide	
3	sulfuric acid and sodium carbonate solution	
4	hydrochloric acid and solid magnesium carbonate	

2 Write balanced full and ionic chemical equations for the reactions in the table in question **1**.

Reaction 1:

Full: ______________________________

Ionic: ______________________________

Reaction 2:

Full: ______________________________

Ionic: ______________________________

Reaction 3:

Full: ______________________________

Ionic: ______________________________

Reaction 4:

Full: ______________________________

Ionic: ______________________________

3 In which of the reactions would you expect to observe bubbles being formed? Explain your answer.

4 In which of the reactions would you expect to observe a change in pH? Explain your answer.

RATING MY LEARNING	My understanding improved	Not confident ◄—► Very confident ○ ○ ○ ○ ○	I answered questions without help	Not confident ◄—► Very confident ○ ○ ○ ○ ○	I corrected my errors without help	Not confident ◄—► Very confident ○ ○ ○ ○ ○

 ISBN 978 1 4886 1933 5

WORKSHEET 3.5

Metals and their cations—reactivity and writing half-equations

A student places a piece of copper metal in a beaker of silver nitrate solution. The solution slowly turns a blue colour.

1 What ion is causing the blue colour in the solution? ______

2 What do you expect the student to observe on the surface of the copper metal?

______

3 Atoms and their ions are species with very different properties. Complete the electron shell diagrams by adding electrons and write each of the words from the box next to the correct diagram to show some different properties of the copper atom and the copper(II) ion.

metal	ion	insoluble	soluble	blue	lustrous

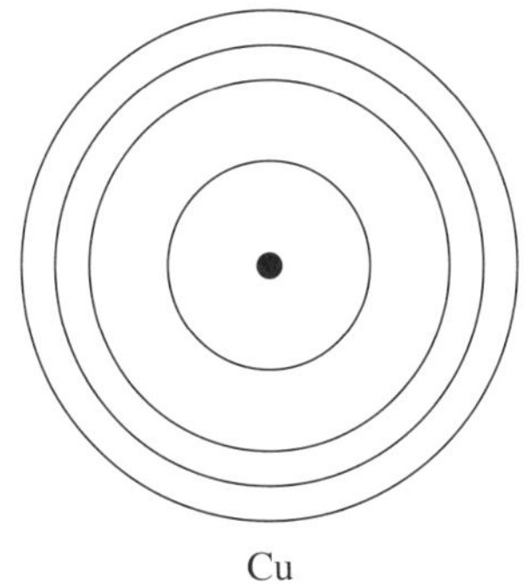

Properties:

______

______

______

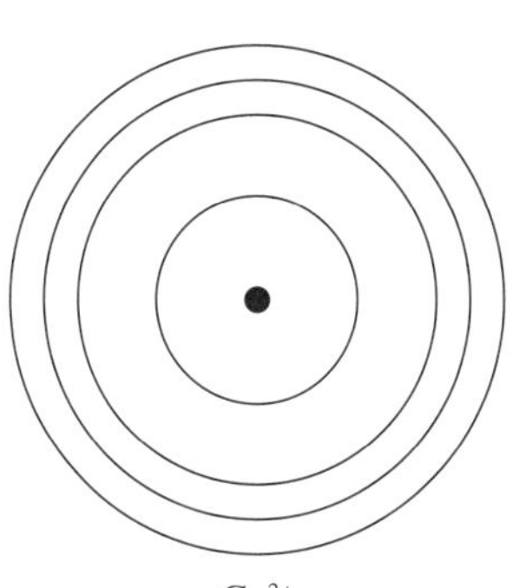

Properties:

______

______

______

4 Complete the half-equations for the oxidation of copper metal and reduction of silver ions by adding one or more electrons where required.

$Cu(s) \rightarrow Cu^{2+}(aq)$

$Ag^{+}(aq) \rightarrow Ag(s)$

5 Write a balanced chemical equation for the overall reaction.

______

ISBN 978 1 4886 1933 5

6 Identify the oxidising agent and reducing agent in the reaction in question **4**.

An activity series of metals can be an ordered listing of reduction half-equations. It can be used to predict redox reactions.

$$K^+(aq) + e^- \rightleftharpoons K(s)$$
$$Na^+(aq) + e^- \rightleftharpoons Na(s)$$
$$Ca^{2+}(aq) + 2e^- \rightleftharpoons Ca(s)$$
$$Mg^{2+}(aq) + 2e^- \rightleftharpoons Mg(s)$$
$$Al^{3+}(aq) + 3e^- \rightleftharpoons Al(s)$$
$$Zn^{2+}(aq) + 2e^- \rightleftharpoons Zn(s)$$
$$Fe^{2+}(aq) + 2e^- \rightleftharpoons Fe(s)$$
$$Ni^{2+}(aq) + 2e^- \rightleftharpoons Ni(s)$$
$$Sn^{2+}(aq) + 2e^- \rightleftharpoons Sn(s)$$
$$Pb^{2+}(aq) + 2e^- \rightleftharpoons Pb(s)$$
$$Cu^{2+}(aq) + 2e^- \rightleftharpoons Cu(s)$$
$$Ag^+(aq) + e^- \rightleftharpoons Ag(s)$$
$$Au^{3+}(aq) + 3e^- \rightleftharpoons Au(s)$$

decreasing reactivity

Use the activity series of metals shown to complete questions **7–11**.

7 Give the formula of the strongest reducing agent.

8 Give the formula of the weakest reducing agent.

9 Give the formula of the strongest oxidising agent.

10 Give the formula of the weakest oxidising agent.

11 Predict whether reactions between the following substances will occur by placing a tick or cross in the table.

Substances	Prediction of whether a reaction will occur (× or ✓)
silver nitrate and tin	
aluminium and nickel sulfate	
lead nitrate and calcium sulfate	
zinc and copper(II) nitrate	
potassium and sodium	

12 Write balanced full and ionic equations for any reaction in question **11** you predicted would occur.

RATING MY LEARNING	My understanding improved	Not confident ◄──► Very confident ○ ○ ○ ○ ○	I answered questions without help	Not confident ◄──► Very confident ○ ○ ○ ○ ○	I corrected my errors without help	Not confident ◄──► Very confident ○ ○ ○ ○ ○

 ISBN 978 1 4886 1933 5

WORKSHEET 3.6

Generating electricity—galvanic cells

1 Complete the concept map summarising the key ideas about galvanic cells using terms from the box. (Some terms may be used more than once, others may not be used.)

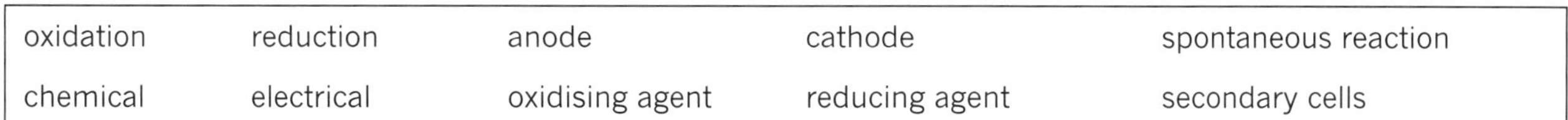

oxidation	reduction	anode	cathode	spontaneous reaction
chemical	electrical	oxidising agent	reducing agent	secondary cells

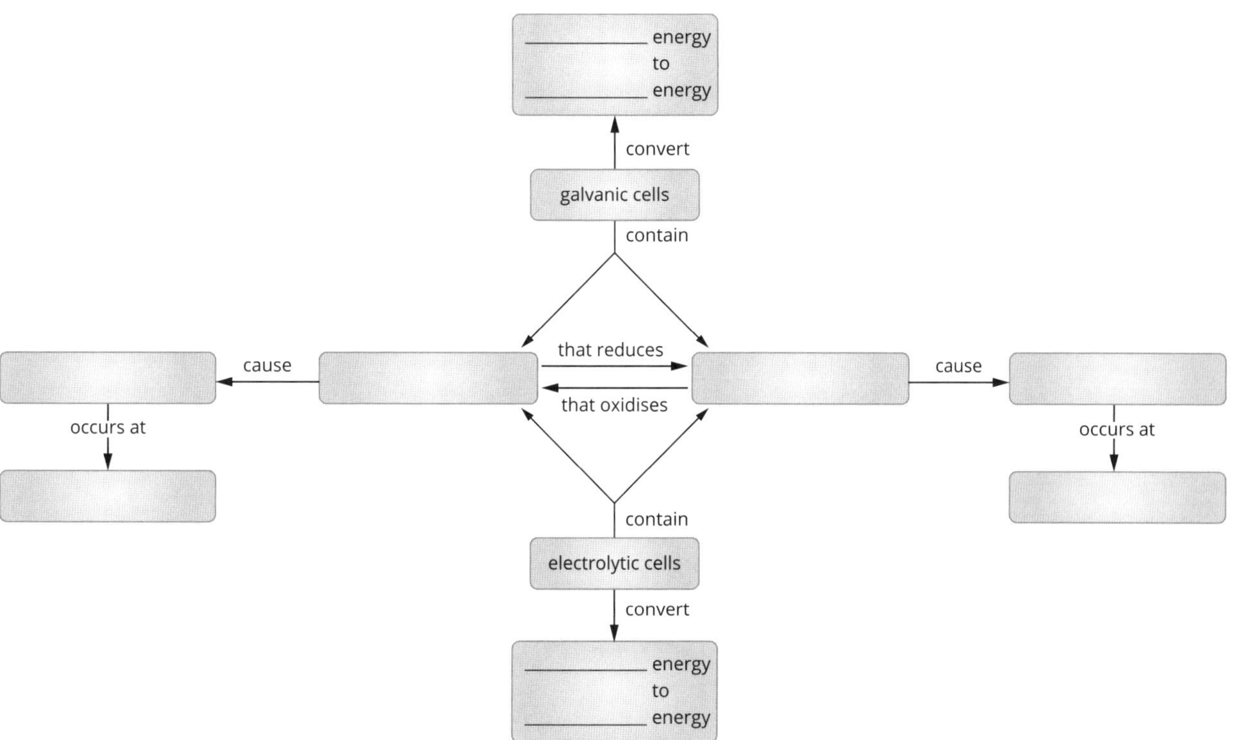

2 Give the meanings of the two terms that were not used in the concept map and use an everyday example to illustrate each term.

Term 1: ______________________ Example: ______________________

Meaning:

____

Term 2: ______________________ Example: ______________________

Meaning:

____

3 The table of standard reduction potentials is used to compare the relative strengths of oxidising agents and reducing agents.

a Draw and label a diagram of the standard reference half-cell used for comparing half-cells. Show the construction of the electrode and the composition of the electrolyte.

b The table of standard reduction potentials shows the relative strengths of oxidising agents and reducing agents at standard conditions. Explain what is meant by 'standard conditions'.

4 a Use the table of standard reduction potentials to predict the reaction that occurs in each of the following galvanic cells, writing the appropriate reactions in the table. The $E°$ for the MnO_4^-/Mn^{2+} half-cell is 1.49 V.

Galvanic cell	Anode reaction	Cathode reaction	Overall cell reaction
Half-cell 1: Cu metal electrode, $CuSO_4(aq)$ Half-cell 2: Mg metal electrode, $MgSO_4(aq)$			
Half-cell 1: $Cl_2(g)$, $Cl^-(aq)$, inert electrode Half-cell 2: $MnO_4^-(aq)$, $Mn^{2+}(aq)$, inert electrode, $H^+(aq)$			
Half-cell 1: Zn metal electrode, $ZnSO_4(aq)$ Half-cell 2: $O_2(g)$, $H^+(aq)$, graphite electrode			

b Suppose that all the contents of the half-cells for one of the cells were mixed together in one container. Give two reasons why the predicted reaction might not be observed.

RATING MY LEARNING	My understanding improved	Not confident ← → Very confident ○ ○ ○ ○ ○	I answered questions without help	Not confident ← → Very confident ○ ○ ○ ○ ○	I corrected my errors without help	Not confident ← → Very confident ○ ○ ○ ○ ○

 ISBN 978 1 4886 1933 5

WORKSHEET 3.7

Reaction routes—rate of reaction

1 Use the terms in the box to help you complete this question. Some terms may be used more than once and others not at all.

collision	frequency	activation	kinetic energy
orientation			

Factors that affect the rate of reaction can be explained in terms of ______________ theory. Reactions occur when particles of the reactants collide with sufficient energy to produce a reaction. This energy needs to be greater than the ______________ energy. Two factors determine whether a reaction occurs. They are the:

- ______________ of the colliding particles
- ______________ of the particles when they collide.

The overall rate of reaction also depends on the ______________ of collisions.

2 The factors that affect the rate of reaction can be summarised. Use the terms in the box to help you complete the diagram. Some terms may be used more than once and others not at all.

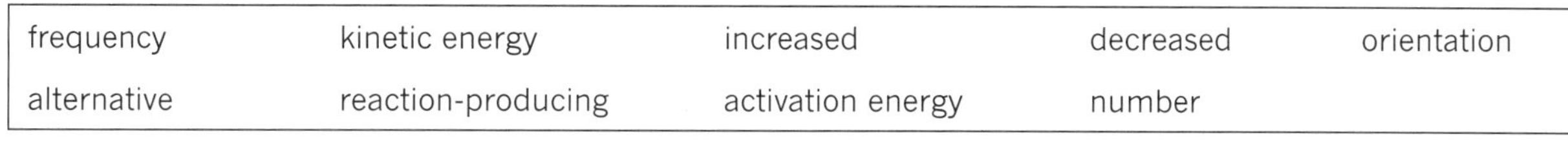

frequency	kinetic energy	increased	decreased	orientation
alternative	reaction-producing	activation energy	number	

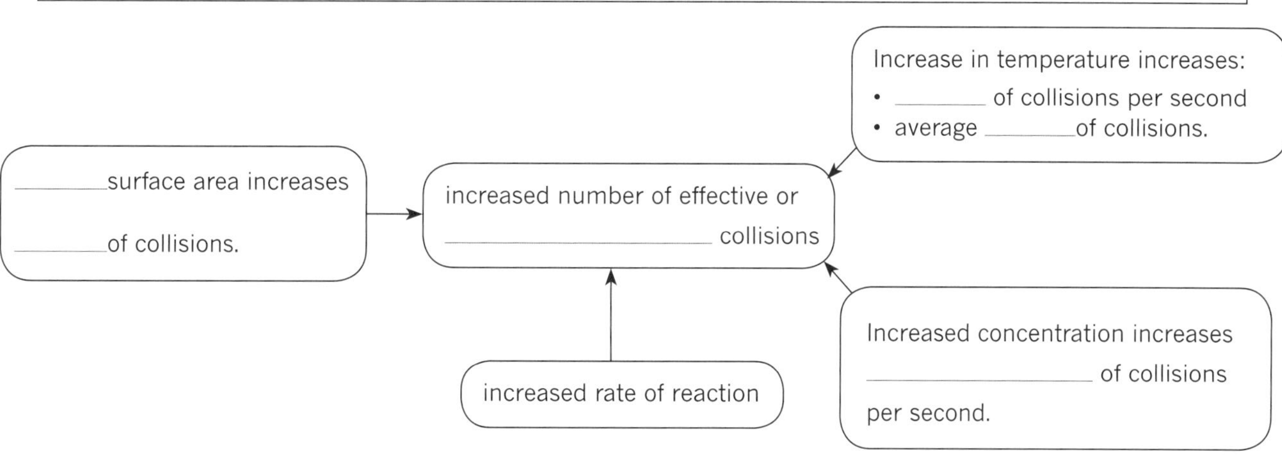

3 The graph shows the effect of increased temperature on the energy of particles in a chemical reaction. The dotted vertical line represents the energy needed to break the bonds in the reactants.

Use the terms in the box to label the diagram.

20°C	30°C	E_a	activation energy	kinetic energy	number of particles

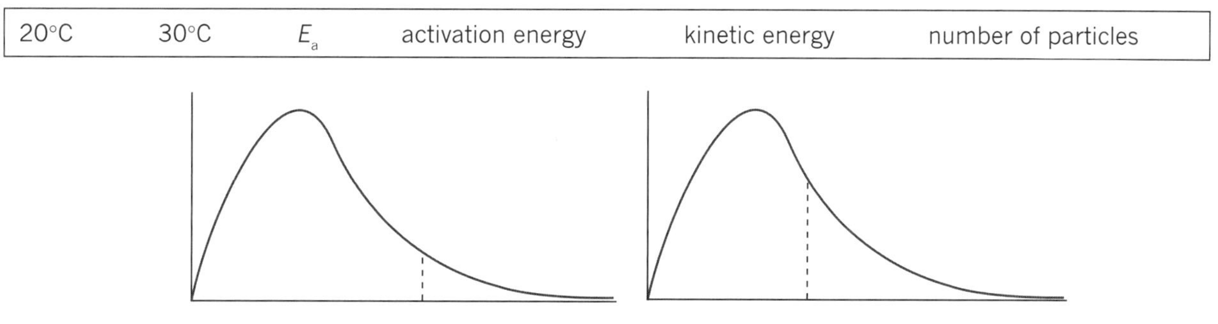

RATING MY LEARNING	My understanding improved	Not confident ◄──► Very confident ○ ○ ○ ○ ○	I answered questions without help	Not confident ◄──► Very confident ○ ○ ○ ○ ○	I corrected my errors without help	Not confident ◄──► Very confident ○ ○ ○ ○ ○

ISBN 978 1 4886 1933 5

WORKSHEET 3.8

Literacy review—full and ionic chemical equations

1 Choose from the terms in the box to complete the summary statements outlining the key points about full and ionic chemical equations.

conserved	solid	aqueous	gaseous	atoms	rearrangement	
soluble	liquid	dissociate	ionic	created	reactant	full
destroyed	spectator	product	change	balanced	dissolved	

a Chemical equations are a way of representing the ____________ of atoms in a chemical reaction.

b Mass is always ____________ in a chemical reaction. Atoms are neither ____________ nor ____________.

c The number and types of atom present before a chemical reaction are also present afterwards. Hence a chemical reaction must be ____________ for each type of atom.

d The state of each ____________ and ____________ is indicated by the following:

i (s) indicates the substance is present in ____________ form.

ii (aq) indicates the substance is ____________, i.e. ____________ in water.

iii (l) indicates the substance is present in ____________ form.

iv (g) indicates the substance is present in ____________ form.

e A ____________ chemical equation includes all of the ____________ present when a chemical reaction takes place. It does not always describe the chemical ____________ accurately because not all atoms present are always involved in chemical change. ____________ ionic compounds ____________ into their ions in solution. The ions that remain unchanged over the course of the reaction are termed ____________ ions.

f An ____________ chemical equation omits spectator ions. It focuses on the actual reaction taking place.

2 Identify the spectator ions in each of the reactions, then write the ionic equation.

a $2NaBr(aq) + PbCl_2(aq) \rightarrow PbBr_2(s) + 2NaCl(aq)$

Spectator ions: ____________________

Ionic equation: ____

b $HCl(aq) + KOH(aq) \rightarrow KCl(aq) + H_2O(l)$

Spectator ions: ____________________

Ionic equation: ____

c $2HCl(aq) + MgO(s) \rightarrow MgCl_2(aq) + H_2O(l)$

Spectator ions: ____________________

Ionic equation: ____

RATING MY LEARNING	My understanding improved	Not confident ◄──► Very confident ○ ○ ○ ○ ○	I answered questions without help	Not confident ◄──► Very confident ○ ○ ○ ○ ○	I corrected my errors without help	Not confident ◄──► Very confident ○ ○ ○ ○ ○

ISBN 978 1 4886 1933 5

WORKSHEET 3.9

Thinking about my learning

On completion of Module 3: Reactive chemistry, you should be able to describe, explain and apply the relevant scientific ideas. You should be able to interpret, analyse and evaluate data.

1 The table shows the areas of key knowledge covered in this module. Reflect on how well you understand each area. Rate your learning by shading the circle that corresponds to your level of understanding for each section. It may be helpful to use colour as a visual representation. For example:

- green—very confident
- orange—in the middle
- red—starting to develop.

Section focus	**Rate my learning**				
	Starting to develop ◄				► Very confident
Chemical reactions—chemical change, synthesis, decomposition and combustion reactions	○	○	○	○	○
Solubility of ionic substances and precipitation reactions	○	○	○	○	○
Reactions of acids with reactive metals and metal carbonates	○	○	○	○	○
Reactions of metals with oxygen, water and dilute acids and metal ions, differing reactivity of metals	○	○	○	○	○
Activity series of metals—half-equations, standard reduction potentials,	○	○	○	○	○
Redox reactions—writing equations, oxidising and reducing agents, oxidation numbers	○	○	○	○	○
Galvanic cells—predicting reactions, anode, cathode, direction of electron flow, salt bridge	○	○	○	○	○
Rates of reaction—ollision theory, measuring rate, effect of surface area, concentration and temperature	○	○	○	○	○

2 Consider the points you have shaded from 'starting to develop' to 'in the middle'. List specific ideas you can identify that were challenging.

____

____

____

____

____

____

3 Write down two different strategies that you will apply to help further your understanding of these ideas.

____

____

____

____

PRACTICAL ACTIVITY 3.1

Modelling types of reactions

Suggested duration: 50 minutes

INTRODUCTION

Modelling is a useful way to show that atoms are rearranged in chemical reactions but are not created or destroyed.

PURPOSE

To model several types of chemical reactions.

MATERIALS

- commercial molecular model building kit or golf-ball-sized lumps of different colours of plasticine

PROCEDURE

1 Write formulae for the reactant and product molecules in Table 1.

2 Using the appropriately coloured 'atoms', either from the model building kit or made of plasticine, construct models of each of the reactant molecules listed in Table 1.

3 Rearrange the atoms in the reactant molecules to produce the products. You may need to make more than one of the reactant molecules to have the right number of atoms needed to make the products.

4 Classify the reactions as synthesis, decomposition, combustion or precipitation.

RESULTS

Note: Ionic compounds and metals do not contain molecules. When you make a model of the formula of an ionic compound you are actually representing the ratio of the elements present in an ionic network.

TABLE 1 Reactants and products of chemical reactions

Reaction number	Reactants	Products	Classification
1	methane + oxygen	carbon dioxide + water	
2	magnesium + chlorine	magnesium chloride	
3	carbon + oxygen	carbon dioxide	
4	calcium carbonate	calcium oxide + carbon dioxide	
5	magnesium + oxygen	magnesium oxide	
6	silver nitrate + magnesium chloride	silver chloride + magnesium nitrate	
7	sodium hydroxide	sodium oxide + water	
8	sodium oxide + carbon dioxide	sodium carbonate	
9	ethanol + oxygen	carbon dioxide + water	
10	potassium hydroxide + magnesium chloride	potassium chloride + magnesium hydroxide	

ISBN 978 1 4886 1933 5

PRACTICAL ACTIVITY 3.1

ANALYSIS OF RESULTS

1 Explain the basis on which you made your classification of each reaction.

2 Which reactions needed more than one of the reactant(s) to make the products? Explain the reason for this.

3 Consider each reaction that you modelled. Complete Table 2 by writing a balanced chemical equation for each reaction. You do not need to include states.

TABLE 2 Balanced chemical equations for reactions

Reaction number	Balanced chemical equation
1	
2	
3	
4	
5	
6	
7	
8	
9	
10	

PRACTICAL ACTIVITY 3.1

CONCLUSION

RATING MY LEARNING	My understanding improved	Not confident ←→ Very confident ○ ○ ○ ○ ○	I answered questions without help	Not confident ←→ Very confident ○ ○ ○ ○ ○	I corrected my errors without help	Not confident ←→ Very confident ○ ○ ○ ○ ○

ISBN 978 1 4886 1933 5

PRACTICAL ACTIVITY 3.2

Precipitation reactions

Suggested duration: 50 minutes

INTRODUCTION

When two clear solutions containing dissolved ionic salts are mixed together, an insoluble product may form and settle out of the mixture. Knowing which ions form precipitates is essential in many industrial processes and in monitoring and maintaining natural waterways.

PURPOSE

- To observe a number of reactions that involve the formation of a precipitate.
- To distinguish between ions forming precipitates and those that are always soluble.
- To write ionic equations to represent the formation of precipitates.

PRE-LAB SAFETY INFORMATION

Material used	Hazard	Control
0.5 mol L^{-1} sodium hydroxide	irritant to skin and eyes	Wear safety glasses, gloves and a laboratory coat.
0.5 mol L^{-1} sodium carbonate	slightly toxic if ingested	Wear safety glasses, gloves and a laboratory coat.
0.1 mol L^{-1} aluminium nitrate	slightly toxic if ingested	Wear safety glasses, gloves and a laboratory coat.
0.1 mol L^{-1} calcium nitrate	slightly toxic if ingested	Wear safety glasses, gloves and a laboratory coat.
0.1 mol L^{-1} copper(II) nitrate	serious irritant to skin and eyes; slightly toxic if ingested	Wear safety glasses, gloves and a laboratory coat.
0.1 mol L^{-1} zinc nitrate	slightly toxic if ingested; toxic to environment	Wear safety glasses, gloves and a laboratory coat. Use appropriate disposal methods.
0.1 mol L^{-1} silver nitrate	serious irritant to skin and eyes; toxic if ingested; stains skin black	Wear safety glasses, gloves and a laboratory coat.
0.1 mol L^{-1} barium nitrate	toxic if ingested	Wear safety glasses, gloves and a laboratory coat.
Product	**Hazard**	**Control**
calcium hydroxide	slightly toxic if ingested	Wear safety glasses, gloves and a laboratory coat.

Please indicate that you have understood the information in the safety table.

Name (print): ______________________________

I understand the safety information (signature): ______________________________

MATERIALS

- dropper bottles, each containing 10 mL of the following 0.5 mol L^{-1} solutions:
  - sodium sulfate, Na_2SO_4
  - sodium chloride, NaCl
  - sodium hydroxide, NaOH
  - sodium carbonate, Na_2CO_3
- dropper bottles, each containing 10 mL of the following 0.1 mol L^{-1} solutions:
  - magnesium nitrate, $Mg(NO_3)_2$
  - calcium nitrate, $Ca(NO_3)_2$
  - copper(II) nitrate, $Cu(NO_3)_2$
  - zinc nitrate, $Zn(NO_3)_2$
  - potassium nitrate, KNO_3
  - silver nitrate, $AgNO_3$
  - barium nitrate, $Ba(NO_3)_2$
- plastic well set or plastic grid
- marker pen
- 5 dropper bottles, randomly labelled A–E, containing 0.1 mol L^{-1} solutions of:
  - potassium nitrate, KNO_3
  - sodium chloride, NaCl
  - sodium sulfate, Na_2SO_4
  - sodium hydroxide, NaOH
  - sodium carbonate, Na_2CO_3
- safety glasses

PRACTICAL ACTIVITY 3.2

Part A—Solubility of ionic compounds

PROCEDURE

Record all of your procedure results in Table 1.

1 Using a marker pen, write the anions you are testing along the top of the plastic wells. Write the cations you are adding down the side of the wells. (Table 1 shows how your grid should look.) Place the well set on a dark background for easier observation.

2 Place two drops of the appropriate cation solution and one drop of the appropriate anion solution into the wells, according to the labels you have written.

3 If no precipitate is formed, record 's' (for soluble).

4 If a precipitate forms, record 'ppt' (for precipitate) and record its colour and quantity (slight, medium or heavy).

RESULTS

TABLE 1 Solubility of ionic compounds

Cation \ Anion	Chloride (Cl^-)	Sulfate (SO_4^{2-})	Hydroxide (OH^-)	Carbonate (CO_3^{2-})
Mg^{2+}				
K^+				
Ca^{2+}				
Cu^{2+}				
Zn^{2+}				
Ag^+				
Ba^{2+}				

ANALYSIS OF RESULTS

1 Which cations:

a did not form any precipitates when they combined with the anions?

b were soluble only as nitrates?

2 Write an ionic equation to represent the formation of each precipitate observed.

 ISBN 978 1 4886 1933 5

PRACTICAL ACTIVITY 3.2

DISCUSSION

3 What generalisations can you make about the solubilities of the following?

a nitrates

b chlorides

c hydroxides

d sulfates

e carbonates

Part B—Identifying unknown solutions

PROCEDURE

Record all of your procedure results in Table 2.

You have been given one of five unknown solutions in dropper bottles. Your dropper bottle may contain a solution of potassium nitrate (KNO_3), sodium chloride (NaCl), sodium sulfate (Na_2SO_4), sodium hydroxide (NaOH) or sodium carbonate (Na_2CO_3).

Select your own set of reagents that will enable you to identify the given unknown sample. Record your results.

Record the label on your unknown sample: ____

TABLE 2 Observations when solutions are mixed

Selected reagent	Observations

ANALYSIS OF RESULTS

1 Write the chemical formula for each precipitate formed in your investigation.

2 Identify your unknown sample. Support your answer with experimental evidence.

ISBN 978 1 4886 1933 5

DISCUSSION

3 It is often possible to convert one salt to another by exploiting solubility differences. Briefly describe, in a series of steps, how you could convert calcium chloride to potassium chloride.

CONCLUSION

RATING MY LEARNING	My understanding improved	Not confident ◄──► Very confident ○ ○ ○ ○ ○	I answered questions without help	Not confident ◄──► Very confident ○ ○ ○ ○ ○	I corrected my errors without help	Not confident ◄──► Very confident ○ ○ ○ ○ ○

PRACTICAL ACTIVITY 3.3

Reactions of HCl with metals and carbonates

Suggested duration: 30 minutes

INTRODUCTION

Acids show similar chemical behaviour. They react with many metals and carbonates in common ways, meaning predictions can be made about the products.

PURPOSE

To predict and observe some reactions of hydrochloric acid with metals and carbonates.

MATERIALS

- 1 mol L^{-1} hydrochloric acid, HCl
- solid samples of calcium carbonate ($CaCO_3$), copper carbonate ($CuCO_3$), sodium hydrogen carbonate ($NaHCO_3$), magnesium and aluminium
- limewater, $Ca(OH)_2$
- taper and matches
- test-tubes/small beakers
- spatula
- teat pipette
- safety glasses

PRE-LAB SAFETY INFORMATION

Material used	Hazard	Control
1 mol L^{-1} hydrochloric acid	toxic by all routes of exposure; lung irritant	Wear safety glasses, gloves and a laboratory coat.
limewater	slightly toxic if ingested; can burn skin and eyes	Wear safety glasses, gloves and a laboratory coat.
calcium carbonate	slightly toxic if ingested	Wear safety glasses, gloves and a laboratory coat.
calcium hydroxide	slightly toxic if ingested	Wear safety glasses, gloves and a laboratory coat.
copper(II) carbonate	serious irritant to skin and eyes; may cause respiratory irritation; slightly toxic if ingested	Wear safety glasses, gloves and a laboratory coat. Avoid breathing in the powder.
magnesium	contact with water releases flammable gases	Wear safety glasses, gloves and a laboratory coat. Do not use near open flame, sparks, etc.
sodium hydrogen carbonate	slightly toxic if ingested	Wear safety glasses, gloves and a laboratory coat.

Please indicate that you have understood the information in the safety table.

Name (print): ____________________

I understand the safety information (signature): ____________________

PROCEDURE

Record all of your procedure results in Table 1.

1 Predict the products of each reaction shown in Table 1 and record the expected products in the appropriate column.

2 Carry out each of the following tests and record your observations. If any gas is evolved, test for hydrogen or carbon dioxide using the two tests shown at the end of this procedure.

- Add about 1 mL of 1 mol L^{-1} HCl to a test-tube containing a small amount of solid $CaCO_3$.
- Add about 1 mL of 1 mol L^{-1} HCl to a small amount of solid $CuCO_3$ in a test-tube. Shake the test-tube and allow it to stand for 5 minutes.
- Add about 10 mL of 1 mol L^{-1} HCl to a test-tube containing a 1 cm strip of magnesium.

ISBN 978 1 4886 1933 5

- Add about 10 mL of 1 mol L^{-1} HCl to a test-tube containing a 1 cm strip of aluminium.
- Add 1 mL of 0.1 mol L^{-1} HCl to a small amount of solid $NaHCO_3$ in a test-tube.
  - *Testing for hydrogen gas:* The pop test is used to test for the presence of hydrogen gas. Evolved gas can be collected in an inverted test-tube and then tested with a lighted taper. If H_2 gas is present, a loud pop will be heard.
  - *Testing for carbon dioxide:* Limewater is used to test for the presence of carbon dioxide. When evolved CO_2 gas is bubbled through limewater, the solution becomes cloudy.

RESULTS

TABLE 1 Predictions and observations for reactions

Reaction	Predictions	Observations
hydrochloric acid + calcium carbonate		
hydrochloric acid + copper(II) carbonate		
hydrochloric acid + magnesium		
hydrochloric acid + aluminium		
hydrochloric acid + sodium hydrogen carbonate		

PRACTICAL ACTIVITY 3.3

ANALYSIS OF RESULTS

1 Comment on how closely your observations match your predictions.

2 Write balanced full and ionic equations for each reaction in Table 2.

TABLE 2 Balanced and full ionic equations for reactions

Reaction	Equations
hydrochloric acid + calcium carbonate	Full: Ionic:
hydrochloric acid + copper(II) carbonate	Full: Ionic:
hydrochloric acid + magnesium	Full: Ionic:
hydrochloric acid + aluminium	Full: Ionic:
hydrochloric acid + sodium hydrogen carbonate	Full: Ionic:

CONCLUSION

RATING MY LEARNING	My understanding improved	Not confident ◄──► Very confident ○ ○ ○ ○ ○	I answered questions without help	Not confident ◄──► Very confident ○ ○ ○ ○ ○	I corrected my errors without help	Not confident ◄──► Very confident ○ ○ ○ ○ ○

ISBN 978 1 4886 1933 5

PRACTICAL ACTIVITY 3.4

Reactivity of metals—student-designed practical activity

Suggested duration: 20 minutes + standing time

INTRODUCTION

Metals have relatively few valence electrons and so they oxidise relatively easily. The relative reactivities of metals can be determined experimentally.

PURPOSE

To compare the reactivity of a variety of metals in water, dilute acid and oxygen.

MATERIALS

- aluminium, small turnings
- small strips of magnesium, iron, copper, zinc
- 1 mol L^{-1} hydrochloric acid, HCl
- safety glasses

Dependent on student design:

- test-tubes and racks
- 50–100 mL beakers
- spatula
- plastic pipette

PRE-LAB SAFETY INFORMATION

Material used	Hazard	Control
1 mol L^{-1} hydrochloric acid	toxic by all routes of exposure; lung irritant	Wear eye and skin protection.
magnesium ribbon	contact with water releases flammable gases	Wear safety glasses, gloves and a laboratory coat. Don't use near open flames, sparks, etc.

Please indicate that you have understood the information in the safety table.

Name (print): ____________________

I understand the safety information (signature): ____________________

PROCEDURE

Use the space to write a method for an experiment that will allow you to compare the reactivity of each of the metals available to you with cold water, hot water, 1 mol L^{-1} HCl and oxygen. When you have designed your method, write out the additional materials you require and have them both checked by your teacher before you proceed.

Additional materials required

____

____

____

____

____

ISBN 978 1 4886 1933 5

RESULTS

Construct a results table to record all of your observations.

ANALYSIS OF RESULTS

1 On the basis of your results, rank the metals in order of their reactivity from most reactive to least reactive.

2 Comment on the suitability of your method. What improvements could you make? How could you extend this investigation?

CONCLUSION

RATING MY LEARNING	My understanding improved	Not confident ◄──► Very confident ○ ○ ○ ○ ○	I answered questions without help	Not confident ◄──► Very confident ○ ○ ○ ○ ○	I corrected my errors without help	Not confident ◄──► Very confident ○ ○ ○ ○ ○

 ISBN 978 1 4886 1933 5

PRACTICAL ACTIVITY 3.5

Activity series of metals

Suggested duration: 40 minutes

INTRODUCTION

In general, ions of metal X in solution will be converted to the metal (that is, displaced from the solution) by any other metal that is more easily oxidised than metal X. The more easily a metal is oxidised, the more reactive it is.

The results of displacement reactions between metals and metallic ions in solution can be used to arrange metals in order of their ability to be oxidised (or the ability of their ions to be reduced) to form a reactivity series.

PURPOSE

To obtain a reactivity series of metals by investigating reactions in which metals displace other metal ions from solutions.

PRE-LAB SAFETY INFORMATION

Material used	Hazard	Control
0.1 mol L^{-1} magnesium nitrate	slightly toxic	Wear safety glasses, gloves and a laboratory coat.
0.1 mol L^{-1} copper(II) sulfate	irritant to eyes and skin; toxic to environment	Wear safety glasses, gloves and a laboratory coat. Use appropriate disposal methods.
0.1 mol L^{-1} iron(II) sulfate	irritant to eyes and skin	Wear safety glasses, gloves and a laboratory coat.
0.1 mol L^{-1} zinc sulfate	causes serious eye damage highly toxic by all routes of exposure; toxic to environment	Wear safety glasses, gloves and a laboratory coat. Use appropriate disposal methods.
magnesium ribbon	contact with water releases flammable gases	Wear safety glasses, gloves and a laboratory coat. Do not use near open flame, sparks/etc.

Please indicate that you have understood the information in the safety table.

Name (print): ____________________

I understand the safety information (signature): ____________________

MATERIALS

- 40 mL 0.1 mol L^{-1} magnesium nitrate solution, $Mg(NO_3)_2$
- 40 mL 0.1 mol L^{-1} copper(II) sulfate solution, $CuSO_4$
- 40 mL 0.1 mol L^{-1} zinc sulfate solution, $ZnSO_4$
- 40 mL 0.1 mol L^{-1} iron(II) sulfate, $FeSO_4$
- 4 pieces of magnesium, approx. 2 cm long
- 4 pieces of copper, approx. 1 cm^2
- 4 pieces of zinc, approx. 1 cm^2
- 4 iron nails or iron wire
- emery paper
- 16 test-tubes
- test-tube rack
- safety glasses

PROCEDURE

1 Clean a sample of each of the four metals using emery paper. Place each sample in a test-tube.

2 Add approximately 5 mL magnesium nitrate solution to each of the test-tubes.

3 Leave them to stand for about 5 minutes. Record any changes that occur.

4 Clean out the test-tubes thoroughly and repeat steps 1–3 using each of the other solutions in turn. Complete Table 1. Draw a cross where there was no observable reaction. Draw a tick and record your observations when there was an observable reaction.

PRACTICAL ACTIVITY 3.5

RESULTS

TABLE 1 Results of reactions between metals and metal ions

Metal \ Metal ion	Mg^{2+}	Cu^{2+}	Zn^{2+}	Fe^{2+}
Mg				
Cu				
Zn				
Fe				

ANALYSIS OF RESULTS

1 a Which metal was the most reactive? ____________________

b Which metal was the least reactive? ____________________

c List the metals in order from most reactive to least reactive.

2 Write half-equations and full equations for the redox reactions that took place during this experiment.

ISBN 978 1 4886 1933 5

3 List the reduction half-equations as a table of standard reduction potentials.

4 Compare the order of your equations with that of the equations listed in the Key knowledge section (Table 3.7). Why might differences arise? GO TO ➤ page 102

CONCLUSION

RATING MY LEARNING	My understanding improved	Not confident ◀——▶ Very confident ○ ○ ○ ○ ○	I answered questions without help	Not confident ◀——▶ Very confident ○ ○ ○ ○ ○	I corrected my errors without help	Not confident ◀——▶ Very confident ○ ○ ○ ○ ○

PRACTICAL ACTIVITY 3.6

Order of half-equations in the table of standard reduction potentials—student-designed practical activity

Suggested duration: 40 minutes for design, 40 minutes for practical activity

INTRODUCTION

During the operation of a galvanic cell, electrons are produced at the negative electrode and flow through the external circuit to the positive electrode. At the negative electrode an oxidation reaction supplies the electrons, whereas at the positive electrode a reduction reaction consumes electrons.

The polarity of the electrodes depends on the relative ability of chemicals at each electrode to donate and accept electrons. The stronger reducing agent undergoes oxidation at the negative electrode.

PURPOSE

To design and perform an experiment to determine the order of four half-equations in the table of standard reduction potentials.

As directed by your teacher, complete the risk assessment and management table by referring to the hazard labels on the reagent bottles or safety data sheets (SDS) or your teacher's risk assessment for the activity.

MATERIALS

- 4 × unidentified half-cells, labelled A–D
- 50 mL 0.1 mol L^{-1} potassium nitrate solution, KNO_3
- 100 mL beaker
- voltmeter
- 2 × wire leads with alligator clips
- 10 × filter papers, 12.5 cm Whatman No. 1
- safety glasses

PRE-LAB SAFETY INFORMATION

Material used	Hazard	Control

Complete and indicate that you have understood the information in the safety table.

Name (print): ____________________

I understand the safety information (signature): ____________________

 ISBN 978 1 4886 1933 5

PROCEDURE

1 Design an experiment that will enable you to determine the order in the table of standard reduction potentials of the half-equations that represent the reactions occurring in four unidentified half-cells, labelled A–D. Clearly indicate how you intend to use your knowledge of galvanic cells and the table of standard reduction potentials to determine the order of the half-equations.

2 Check the design of your experiment with your teacher.

3 Perform the experiment and tabulate your results in your logbook or in the space provided.

RESULTS

Tabulate your results in your log book or in this space.

ANALYSIS OF RESULTS

1 In light of your results, determine the order of the four half-equations as they would appear in the table of standard reduction potentials.

2 Comment on the success or otherwise of your experiment. How could you improve your experiment?

3 Comment on the reliability of conclusions you can draw from this investigation.

CONCLUSION

RATING MY LEARNING	My understanding improved	Not confident ←→ Very confident ○ ○ ○ ○ ○	I answered questions without help	Not confident ←→ Very confident ○ ○ ○ ○ ○	I corrected my errors without help	Not confident ←→ Very confident ○ ○ ○ ○ ○

 ISBN 978 1 4886 1933 5

PRACTICAL ACTIVITY 3.7

Factors affecting rate of reaction

Suggested duration: 40 minutes

INTRODUCTION

Hydrochloric acid reacts with sodium thiosulfate to produce sulfur, sulfur dioxide gas and water:

$$S_2O_3^{2-}(aq) + 2H^+(aq) \rightarrow S(s) + SO_2(g) + H_2O(l)$$

The sulfur precipitate causes the reaction mixture to become cloudy. The rate at which the solution becomes cloudy indicates the rate of reaction: the shorter the period before the cloudiness obscures a cross on a sheet of paper placed under the beaker, the more rapid the rate of reaction. In this practical activity the effects of temperature and concentration of one of the reactants on the rate of reaction are studied.

PURPOSE

To investigate the ways in which changes in temperature and concentration of reactants affect the rate of a reaction.

As directed by your teacher, complete the risk assessment and management table by referring to the hazard labels on the reagent bottles or safety data sheets (SDS) or your teacher's risk assessment for the activity.

MATERIALS

- 40 mL 2 mol L^{-1} hydrochloric acid, HCl
- 100 mL 0.25 mol L^{-1} sodium thiosulfate solution, $S_2O_3^{2-}$
- 10 mL measuring cylinder
- 100 mL measuring cylinder
- 6 × 100 mL beakers
- thermometer, −10 to 110°C
- Bunsen burner, tripod stand and gauze mat
- bench mat
- stopwatch
- black marking pen
- safety glasses

PRE-LAB SAFETY INFORMATION

Material used	Hazard	Control
sulfur dioxide gas	toxic by inhalation; causes burns; risk of serious damage to the eyes	Avoid breathing gas: work in a well-ventilated area and dispose of reaction mixtures using a sink in a fume cupboard. Wear safety glasses.
2 mol L^{-1} hydrochloric acid		
0.25 mol L^{-1} sodium thiosulfate		

Complete and indicate that you have understood the information in the safety table.

Name (print): ______________________________

I understand the safety information (signature): ______________________________

PROCEDURE

Part A—Effect of temperature

Record all of your procedure results in Table 1.

1. Mark a cross on a sheet of paper with a black marking pen.
2. Place a 100 mL beaker on top of the cross. Pour 10 mL of 0.25 mol L^{-1} $Na_2S_2O_3$ solution and 35 mL water into the beaker.
3. Record the temperature of the solution.
4. Add 5 mL of 2 mol L^{-1} HCl and commence timing. Measure and record in the table the time taken for the cross to disappear when viewed from above the beaker.
5. Pour 10 mL of 0.25 mol L^{-1} $Na_2S_2O_3$ solution and 35 mL water into another beaker. Heat the solution to 40°C. Remove the beaker from the source of heat, place it on top of the cross, and immediately add 5 mL HCl. Again, measure and record the time taken for the cross to disappear.

6 Repeat step 5 with a third beaker, this time heating the solution to 60°C.

Part B—Effect of concentration of reactants

Record all of your procedure results in Table 2.

1 Place a 100 mL beaker on top of a cross marked on a sheet of paper. Pour 10 mL of 0.25 mol L^{-1} $Na_2S_2O_3$ solution and 35 mL water into the beaker. Add 5 mL of 2 mol L^{-1} HCl and commence timing. Measure and record the time taken for the cross to disappear when viewed from above the beaker.

2 Place another 100 mL beaker on top of the cross. Pour 25 mL of the $Na_2S_2O_3$ solution and 20 mL water into the beaker. Add 5 mL HCl and commence timing. Measure and record the time taken for the cross to disappear.

3 Place another 100 mL beaker on top of the cross. Pour 40 mL sodium thiosulfate solution and 5 mL water into the beaker. Add 5 mL HCl and commence timing. Measure and record the time taken for the cross to disappear.

RESULTS

TABLE 1 Results for effect of temperature

Temperature (°C)	Time for the cross to disappear (s)

TABLE 2 Results for effect of concentration of reactants

Solution	Concentration of sodium thiosulfate in solution (mol L^{-1})	Time for the cross to disappear (s)
10 mL $Na_2S_2O_3$ solution, 35 mL water and 5 mL HCl		
25 mL $Na_2S_2O_3$ solution, 20 mL water and 5 mL HCl		
40 mL $Na_2S_2O_3$ solution, 5 mL water and 5 mL HCl		

ANALYSIS OF RESULTS

1 What effect does an increase in temperature have on the rate of this reaction?

2 Explain why an increase in temperature leads to a change in the rate of reaction.

3 Calculate the concentrations of the sodium thiosulfate in each of the three reaction mixtures made up in part B and enter these in Table 2.

4 What effect does increasing the concentration of sodium thiosulfate have on the rate of the reaction?

 ISBN 978 1 4886 1933 5

DISCUSSION

5 Explain why the rate of a reaction depends upon the concentration of the reactants.

CONCLUSION

RATING MY LEARNING	My understanding improved	Not confident ⟷ Very confident ○ ○ ○ ○ ○	I answered questions without help	Not confident ⟷ Very confident ○ ○ ○ ○ ○	I corrected my errors without help	Not confident ⟷ Very confident ○ ○ ○ ○ ○

DEPTH STUDY 3.1

Reaction rates—practical investigation

Suggested duration: 3.5 hours (including writing time and preparation of poster)

INTRODUCTION

This depth study requires you to design and conduct a quantitative practical investigation related to a factor that affects the rate of a chemical reaction between hydrochloric acid and solid calcium carbonate. You will collect and process data and draw conclusions based on your evidence. You will communicate your results in a scientific poster.

Your investigation must use the reaction between hydrochloric acid and solid calcium carbonate.

PLANNING YOUR INVESTIGATION

The following tasks will help you to design your investigation.

1 **a** Write an equation for the reaction between hydrochloric acid and calcium carbonate.

b Briefly describe how you might be able to track the rate of this chemical reaction.

2 **a** Choose a single factor that will affect the rate of this reaction that you will investigate.

b Write an inquiry question that will be the focus of your investigation.

c Complete the table to identify the variables that are relevant to your inquiry question.

Independent variable	Dependent variable	Controlled variables

d Complete the table to write a hypothesis as well as a clear and relevant aim from the hypothesis and variables.

Hypothesis	Aim

ISBN 978 1 4886 1933 5

e List any background concepts that will be relevant to your investigation. Highlight any key terms you think you should define.

f In the space provided, sketch a table you might include in the scientific poster to present the raw data collected in the investigation.

g Make some suggestions for how you might process the raw data to enable any trends, patterns or scientific relationships to be evident.

METHOD

Use the space provided to write the method for your investigation.

ISBN 978 1 4886 1933 5

CONDUCTING YOUR INVESTIGATION

Use the space provided to construct a table to record your observations and results.

COMMUNICATING—PRESENTING YOUR FINDINGS IN A POSTER

The following subheadings will give you an idea of what you should include in your scientific poster.

Title

- Is your inquiry question original, clear, concise?

Introduction

- Have you identified and defined relevant background concepts?
- Have you identified and justified the dependent, independent and all controlled variables?

Aim

- Is your aim clear?
- Is it relevant to the inquiry question?
- Does it relate directly to the dependent and independent variables?

Hypothesis

- Is your hypothesis a prediction?
- Is it based on the effect of changing one variable on another?

 ISBN 978 1 4886 1933 5

Variables

- What is your independent variable?
- What is your dependent variable?
- What are the variables you will need to control?

Methodology

- Have you comprehensively selected appropriate equipment and materials?
- Does your method comprehensively meet your aim?
- Have you identified, explained and managed risks?
- Have you incorporated safe and ethical work practices?
- Is your method repeatable?

Results

- Have you collected relevant data systematically?
- Is your raw data clear, accurate and detailed?
- Have you processed your data in the most appropriate ways to illustrate trends, patterns and/or scientific relationships?
- Have you comprehensively explained how accuracy, precision, reliability, validity, uncertainty and errors relate to your collected data?

Discussion

- Have you analysed and explained your investigation data?
- Does your discussion address all possible sources of error and uncertainty?
- How is your analysis limited by the data you did or did not collect?
- Were the procedures you used appropriate and effective?
- Have you made practical suggestions for improvements to the experimental design?
- Have you suggested possible further investigations?
- Have you explained the link between your investigation findings and the relevant scientific concepts you included in the Introduction section?
- Have you used scientific language and conventions?

Conclusion

- Is it valid given your analysis and evaluation?
- Does it relate collected data to the hypothesis and/or aim?
- Does it provide a justified response to the investigation question?

References and acknowledgements

- Have you used a standard scientific referencing system?
- Have you referred to specific pages/sections in your logbook?

Poster template

A suggested layout for your poster is shown here with more prompts for what should be in each section. You can access poster design layouts from the internet.

Title *(inquiry question, ending in a '?')*
Student name
School name
Unit 2 Chemistry, *year*

Introduction

Cut and paste from Word document/logbook

- *Include aim, hypothesis*
- *Summarise background chemical concepts, include in-text citation to appropriate reference*
- *Font should be easy to read from 1 metre away (suggest minimum 16 point in Calibri, Arial or Tahoma)*
- *Use contrast of colour (e.g. black or blue colour on white background)*

Methodology

Cut and paste from Word document/logbook

- *Include summary of method used:*
  - *e.g. flowchart*
  - *Consider using SmartArt (in PowerPoint/Word)*
- *Identify and manage relevant risks and follow relevant OH&S guidelines*

Cut and paste images (e.g. from logbook)

Where possible include a labelled diagram of experimental set-up/results

Each figure should be numbered with an appropriate label (and reference where applicable)

Results

Cut and paste from Word document/logbook

- *Include summary of results obtained*
- *Avoid including all raw and processed experimental data—use extract of raw and processed experimental data*
- *Try to summarise data in tables/ graphically*
- *Number each table/graph and give it an appropriate title*
- *Where possible present data graphically (e.g. in pie/bar/line graphs)*
- *Ensure axes are labelled including units if applicable*

Discussion

Cut and paste from Word document/logbook

- *Discuss, analyse and evaluate data obtained (directly refer to data in tables)*
- *Link results to key chemical concepts (include in-text citations as appropriate)*
- *Identify outliers*
- *Include limitations and recommendations for further research*

Conclusion

Cut and paste from Word document/logbook

- *Summarise research findings according to aim and hypothesis*
- *Include recommendations for future research*

References

Include alphabetical listing (by author last name/institution) of all sourced material (ensure in-text citations included where appropriate in poster)

Use consistent referencing style (e.g. APA)

Acknowledgements

 ISBN 978 1 4886 1933 5

Multiple choice

1 What type of chemical reaction is represented by the following equation?

$CaO(s) + CO_2(g) \rightarrow CaCO_3(s)$

A decomposition
B synthesis
C precipitation
D combustion

2 What will be the likely products when glucose, $C_6H_{12}O_6$, undergoes complete combustion in oxygen?

A CO_2 and H_2
B CO, CO_2 and H_2O
C CO, CH_4 and O_2
D CO_2 and H_2O

3 What will be the likely products of the decomposition of nitric acid, HNO_3?

A $NO_3 + H_2O + CO_2$
B $NO_2 + H_2 + O_2$
C $NO_2 + H_2O + O_2$
D $NO_3 + H_2 + CO_2$

4 Which of the following ionic compounds is insoluble in water?

A magnesium sulfate
B iron nitrate
C ammonium hydroxide
D magnesium carbonate

5 Which of the following will undergo a spontaneous reaction when added to sodium nitrate solution?

A zinc nitrate solution
B calcium sulfate powder
C potassium metal
D zinc metal

6 Which of the following combinations of half-cells is most likely to produce a voltage of 0.30 V when connected in a galvanic cell?

A $Mg^{2+}(aq)/Mg(s)$ and $Al^{3+}(aq)/Al(s)$
B $Zn^{2+}(aq)/Zn(s)$ and $Al^{3+}(aq)/Al(s)$
C $Sn^{2+}(aq)/Sn(s)$ and $Fe^{2+}(aq)/Fe(s)$
D $Sn^{2+}(aq)/Sn(s)$ and $Zn^{2+}(aq)/Zn(s)$

7 In which of the following species does nitrogen have the highest oxidation number?

A N_2
B NO_2
C NH_3
D HNO_3

8 Which of the following factors increases the rate of a chemical reaction by increasing the proportion of collisions that are successful?

A increasing the concentration of one reactant
B increasing the temperature of one reactant
C increasing the surface area
D decreasing the temperature of both reactants

Short answer

9 Write a full balanced chemical equation to represent each of the following reactions:

a A precipitation reaction occurs between magnesium chloride solution and silver nitrate solution.

____

b Solid magnesium reacts with hydrochloric acid.

____

c Nitric acid is added to potassium carbonate solution.

____

10 A student investigates the reactivity of metals A, B, C and D by combining the different metals with metal ions in solution and observing whether a reaction occurred. The results of the student's investigation are shown in the table. A tick indicates a reaction was observed. A cross indicates there was no reaction.

Metal ion \ Metal	A	B	C	D
A^+		✗	✓	✗
B^{2+}	✓		✓	✗
C^{3+}	✗	✗		✗
D^+	✓	✓	✓	

ISBN 978 1 4886 1933 5

a Place the metals in order of reactivity from most reactive to least reactive.

b Write a half-equation for the reaction of the strongest oxidant.

c Identify the equation you wrote in part b as oxidation or reduction.

11 A student sets up a galvanic cell using a Cu/Cu^{2+} and Ag/Ag^{+} half-cells.

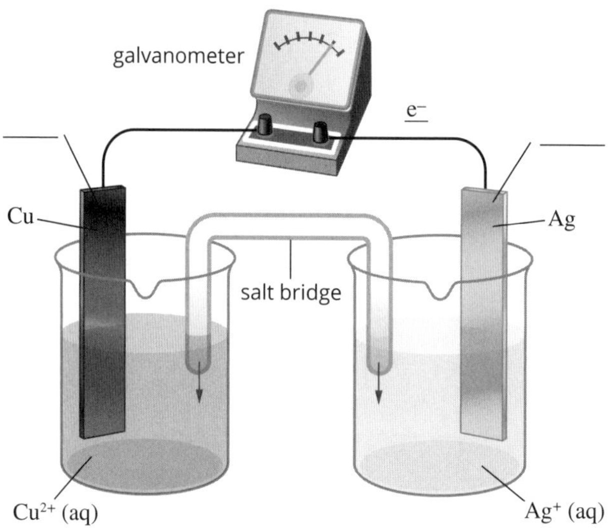

a Label the anode and the cathode.

b Label the negative and positive electrodes.

c Indicate the direction of electron flow with an arrow.

d Write half-equations for the reactions that occur at the:

i anode

ii cathode.

e Write an overall equation for the reaction that occurs in the galvanic cell.

f What is the voltage that would be produced by this cell?

g Suggest a suitable solution that could be used in the salt bridge.

 ISBN 978 1 4886 1933 5

Extended response

12 A student investigating the effect of temperature on reaction rates sets up three different reactions between hydrochloric acid and calcium carbonate. The reaction conditions and results are shown in the table. In the first two reactions the student uses 1.0 g of large calcium carbonate chips; however, they run out. The student uses smaller chips in the third reaction. The student monitors the reaction rate by carrying out the reaction on a balance and measuring the mass loss at 1 minute and 2 minutes.

	Reaction 1	**Reaction 2**	**Reaction 3**
mass of calcium carbonate	1.0 g large chips	1.0 g large chips	1.0 g smaller chips
temperature of 0.1 mol L^{-1} HCl	10°C	20°C	30°C
mass of flask and reactants at the start	235 g	235 g	235 g
mass of flask and reactants after 1 minute	234.2 g	233.5 g	231.5 g
mass of flask and reactants after 2 minutes	233.6 g	232.7 g	231.4 g

a Suggest three improvements to this practical investigation that would increase the validity of conclusions.

b The experiment appears to indicate that increasing the temperature of the HCl does increase the rate of the reaction.

i Which observations best support this?

ii How valid is this conclusion?

iii Explain the theory behind why increasing temperature will increase the rate of a reaction.

MODULE 4 Drivers of reactions

Outcome

By the end of this module you will be able to:

- develop and evaluate questions and hypotheses for scientific investigation (CH11-1)
- analyse and evaluate primary and secondary data and information (CH11-5)
- solve scientific problems using primary and secondary data, critical thinking skills and scientific processes (CH11-6)
- communicate scientific understanding using suitable language and terminology for a specific audience or purpose (CH11-7)
- analyse the energy considerations in the driving force for chemical reactions (CH11-11)

Content

ENERGY CHANGES IN CHEMICAL REACTIONS

INQUIRY QUESTION **What energy changes occur in chemical reactions?**

By the end of this module you will be able to:

- conduct a practical investigation, using appropriate tools (including digital technologies), to collect data, analyse and report on how the rate of a chemical reaction can be affected by a range of factors, including but not limited to: ICT N
  - catalysts (ACSCH042)
- conduct practical investigations to measure temperature changes in examples of endothermic and exothermic reactions, including: CCT ICT L
  - combustion
  - dissociation of ionic substances in aqueous solution (ACSCH018, ACSCH037) N
- investigate enthalpy changes in reactions using calorimetry and $q = mc\Delta T$ (heat capacity formula) to calculate, analyse and compare experimental results with reliable secondary-sourced data, and to explain any differences CCT ICT N
- construct energy profile diagrams to represent and analyse the enthalpy changes and activation energy associated with a chemical reaction (ACSCH072) ICT
- model and analyse the role of catalysts in reactions (ACSCH073) ICT

ENTHALPY AND HESS'S LAW

INQUIRY QUESTION **How much energy does it take to break bonds, and how much is released when bonds are formed?**

By the end of this module you will be able to:

- explain the enthalpy changes in a reaction in terms of breaking and reforming bonds, and relate this to: L
  - the law of conservation of energy

Module 4 • Drivers of reactions

- investigate Hess's law in quantifying the enthalpy change for a stepped reaction using standard enthalpy change data and bond energy data, for example: (ACSCH037) CCT ICT N
  - carbon reacting with oxygen to form carbon dioxide via carbon monoxide
- apply Hess's law to simple energy cycles and solve problems to quantify enthalpy changes within reactions, including but not limited to: ICT N
  - heat of combustion
  - enthalpy changes involved in photosynthesis
  - enthalpy changes involved in respiration (ACSCH037)

ENTROPY AND GIBBS FREE ENERGY

INQUIRY QUESTION **How can enthalpy and entropy be used to explain reaction spontaneity?**

By the end of this module you will be able to:

- analyse the differences between entropy and enthalpy L
- use modelling to illustrate entropy changes in reactions CCT ICT
- predict entropy changes from balanced chemical reactions to classify as increasing or decreasing entropy CCT
- explain reaction spontaneity using terminology, including: (ACSCH072) L
  - Gibbs free energy
  - enthalpy
  - entropy
- solve problems using standard references and $\Delta G^\circ = \Delta H^\circ - T\Delta S^\circ$ (Gibbs free energy formula) to classify reactions as spontaneous or nonspontaneous ICT N
  - predict the effect of temperature changes on spontaneity (ACSCH070) CCT

Key knowledge

Energy changes in chemical reactions

Chemical energy is stored in the bonds between atoms and between molecules. It results from attractions and repulsions between electrons and protons, and vibrations and rotations of nuclei. The unit for energy is the **joule**, J. Energy is also often expressed in kilojoules, kJ. 1 kJ = 1000 J.

The **chemical energy** of a substance is called its **heat content** or **enthalpy** and is given the symbol *H*. We can measure the change in enthalpy, ΔH, for a chemical reaction:

$$\Delta H = H_{\text{products}} - H_{\text{reactants}}$$

EXOTHERMIC AND ENDOTHERMIC REACTIONS

Depending on the enthalpy change that occurs, a reaction is described as **exothermic** or **endothermic** (Table 4.1).

TABLE 4.1 Comparison of exothermic and endothermic reactions

Exothermic reaction	Endothermic reaction
More energy is released than absorbed (Figure 4.1a).	More energy is absorbed than released (Figure 4.1b).
ΔH is negative	ΔH is positive

During a chemical reaction the bonds of the reactant particles absorb energy and break (Figure 4.1). The new bonds of the product particles are then formed, releasing energy. The energy needed for a reaction to occur is called the **activation energy** (Figure 4.2).

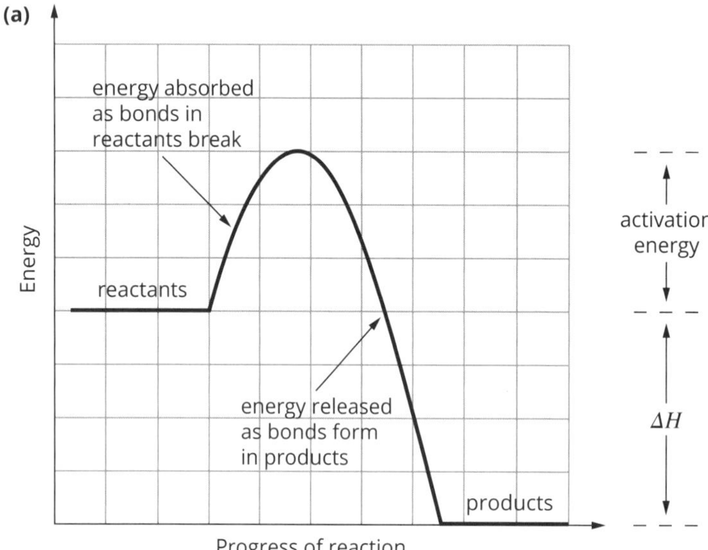

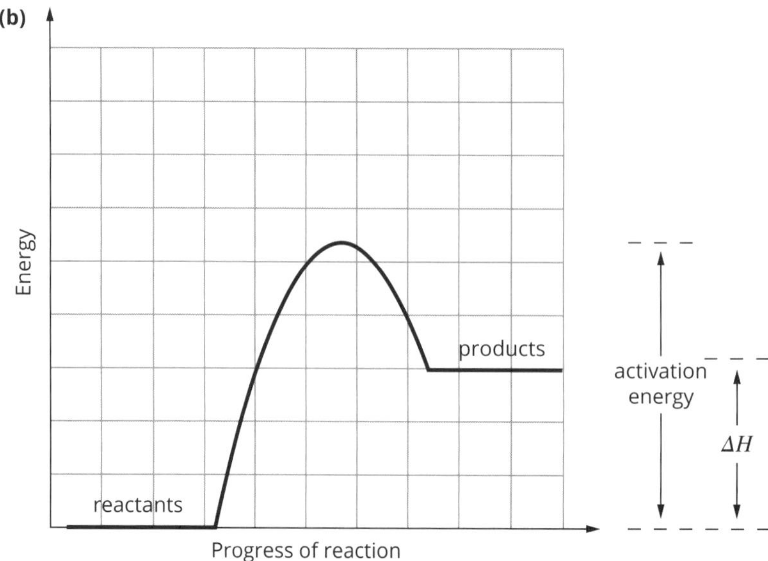

FIGURE 4.1 Energy profile diagrams of (a) exothermic and (b) endothermic reactions

ISBN 978 1 4886 1933 5

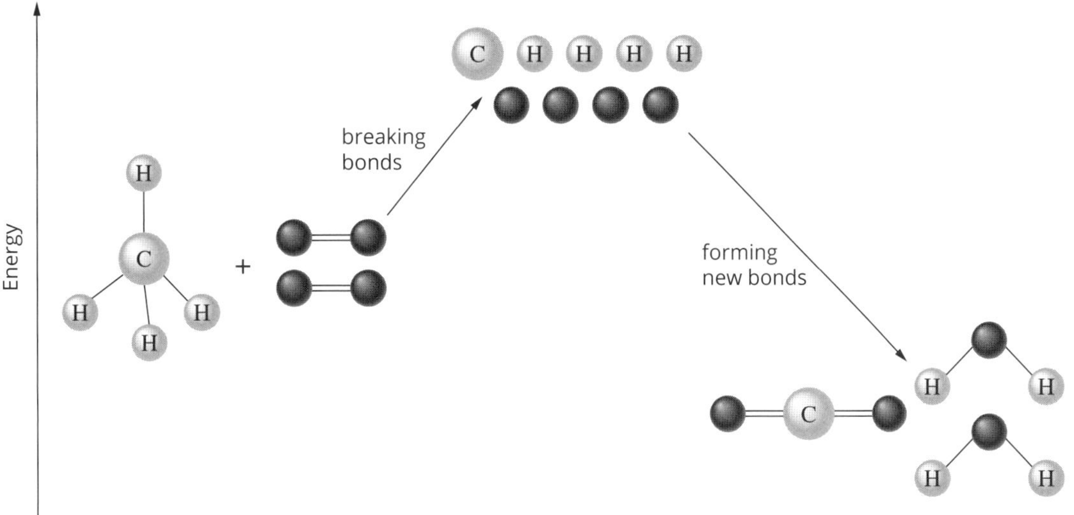

FIGURE 4.2 Energy changes in the breaking and forming of bonds in the reaction between methane and oxygen

The activation energy barrier is evident in the energy profile diagrams in both endothermic and exothermic reactions. Activation energy is the energy that must be absorbed to break the bonds in the reactants to allow the reaction to proceed.

Combustion reactions are exothermic reactions that release a large amount of energy.

The dissociation of ionic substances in aqueous solutions are endothermic reactions that absorb energy from their surroundings.

THERMOCHEMICAL EQUATIONS

Thermochemical equations are chemical equations that include a ΔH value, showing the energy released or absorbed if the moles of reactants shown in the equation react completely. For example, when two moles of octane (C_8H_{18}) are burnt, 10 108 kJ of energy is released:

$$2C_8H_{18}(l) + 25O_2(g) \rightarrow 16CO_2(g) + 18H_2O(l)$$
$$\Delta H = -10\,108\,kJ\,mol^{-1}$$

If one mole of octane is burnt, only 5054 kJ of energy is released. When the equation is reversed, the sign of the ΔH is also reversed:

$$16CO_2(g) + 18H_2O(l) \rightarrow 2C_8H_{18}(l) + 25O_2(g)$$
$$\Delta H = +10\,108\,kJ\,mol^{-1}$$

For thermochemical equations, it is important to understand that:

- when an equation is reversed, the ΔH for the reversed reaction has the same magnitude but opposite sign from the original equation
- when the coefficients in an equation are doubled, the ΔH for the new equation will be doubled; similarly, if the coefficients in an equation are halved, the ΔH for the new equation will be halved
- the states of the reactants and products are important because a change of state involves a change in enthalpy.

The **specific heat capacity** of a substance is a measure of the amount of energy (usually in J) needed to increase the temperature of a specific quantity of that substance (usually 1 g) by 1 K.

HEAT OF COMBUSTION

The **heat of combustion** is the enthalpy change that occurs when a specified amount (e.g. 1 mol) of a substance burns completely in oxygen. It is usually measured at conditions of 298 K and 100 kPa. The heat of combustion is given the symbol ΔH_c and is measured in $kJ\,mol^{-1}$ when it is a **pure substance**. A pure substance is a substance comprising only the one type of substance, not a mixture. If a substance is a mixture, then $kJ\,g^{-1}$ or $kJ\,L^{-1}$ are used.

Determining the heat of combustion

An estimate of the heat of combustion of a substance can be obtained using the equipment shown in Figure 4.3.

The energy transferred to the water can be calculated by measuring the volume of water and the change in temperature. The energy is calculated using the formula:

$$q = m \times c \times \Delta T$$

where:

- q is the energy that is transferred to the water in joules (J)
- m is the mass of the water in grams (g), which is equal to the volume of water in millilitres (mL)
- c is the specific heat capacity of the water ($4.18\,J\,g^{-1}\,K^{-1}$)
- ΔT is the change in temperature in kelvin (K).

The values obtained by this method will not be the same as the theoretical values for the fuel because only some of the heat released in the combustion reaction will be absorbed by the water, producing a smaller ΔT and therefore a lower estimate of the energy content.

Calorimetry

Calorimetry is a technique for measuring the energy change during both exothermic and endothermic chemical reactions. There are two types of calorimeter (Figure 4.4):

- bomb calorimeters—for reactions that involve gaseous reactants or products. Bomb calorimeters are used to measure energy changes that occur during the combustion of fuels or food

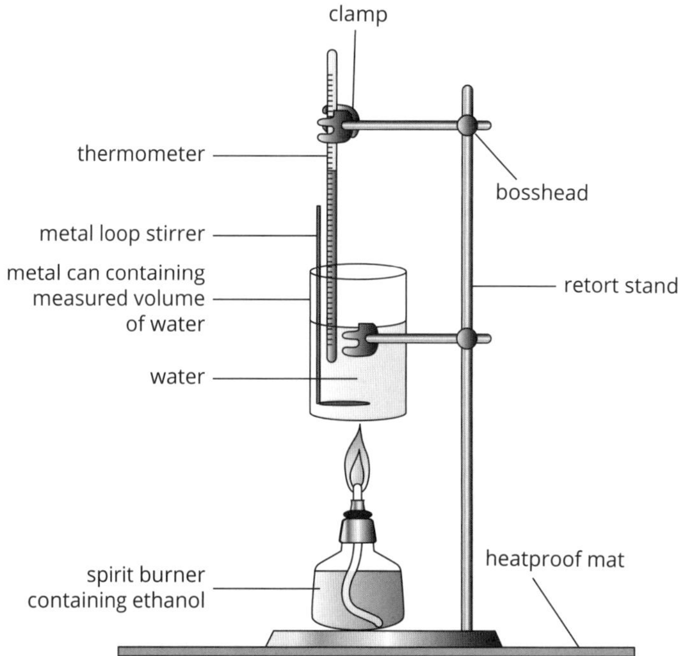

FIGURE 4.3 Apparatus for measuring heat of combustion of a fuel (for example, ethanol). A metal can containing a measured volume of water is held above the wick of a spirit burner.

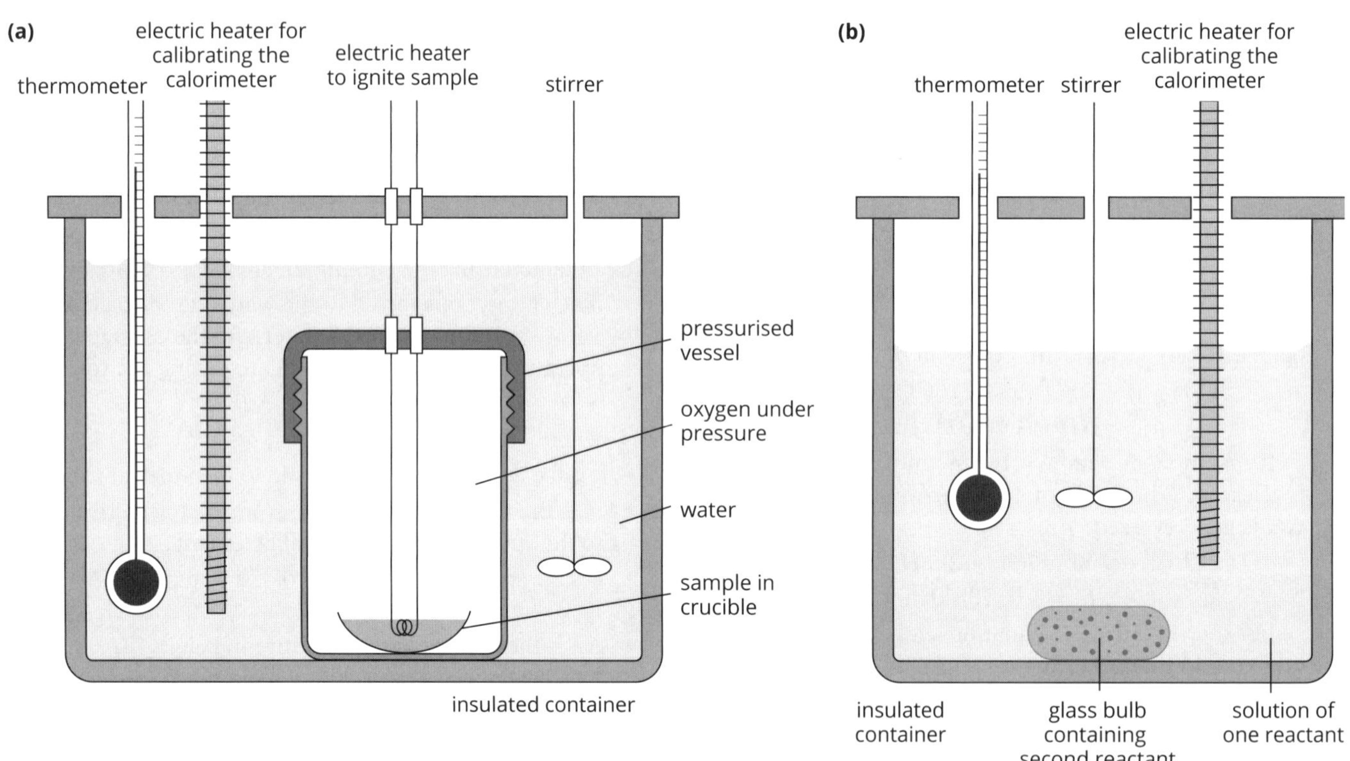

FIGURE 4.4 (a) A bomb calorimeter and (b) a solution calorimeter are used for measuring energy changes in reactions.

- solution calorimeters—for reactions in solution.

Both types of calorimeter have a large, insulated reservoir of water. The heat released or absorbed by a chemical reaction within the calorimeter causes the temperature of the water to change and this temperature change is measured.

ENTHALPY OF DISSOLUTION

Solutes dissolve in solvents to form solutions (Figure 4.5). This process is called dissolution. The ability of a solvent to dissolve requires an adequate force of attraction between the solute and solvent particles.

The polar nature of water molecules and the presence of hydrogen bonds enable water to dissolve other polar covalent substances as well as many ionic substances. Water does not readily dissolve non-polar substances.

ISBN 978 1 4886 1933 5

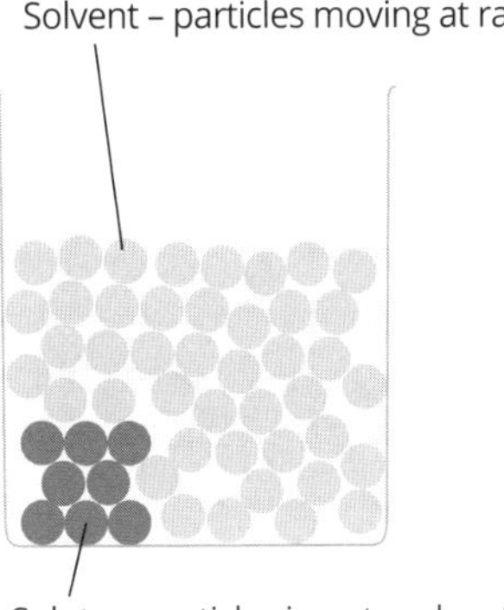

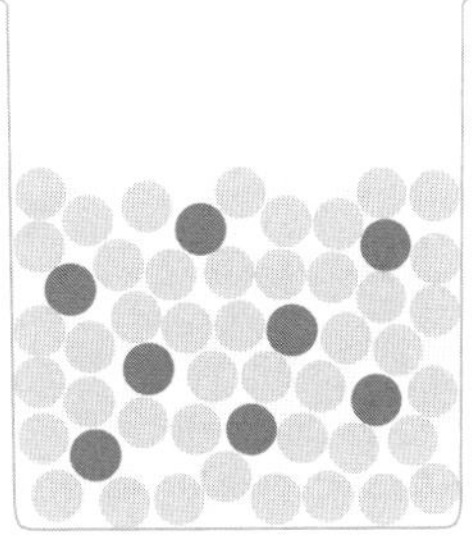

FIGURE 4.5 Rearrangement of particles when a solute dissolves in a solvent

Many ionic compounds dissolve readily in water. Attraction to the polar water molecules causes the ions in some ionic substances to **dissociate** from each other and become hydrated. An example is sodium chloride:

$$MgCl_2(s) \xrightarrow{H_2O(l)} Mg^{2+}(aq) + 2Cl^-(aq)$$

During this process there is an enthalpy change. The temperature change can be measured using a solution calorimeter and the enthalpy change calculated using the calibration factor of the calorimeter as already outlined.

CATALYSTS

A catalyst is a factor that can increase the rate of a chemical reaction. The effect of a catalyst is shown in the energy profile diagrams in Figure 4.6. Catalysts:

- provide an alternative reaction pathway, which has a lower activation energy
- do not change the ΔH for the reaction
- are not consumed during the reactions they speed up
- need only be present in small quantities.

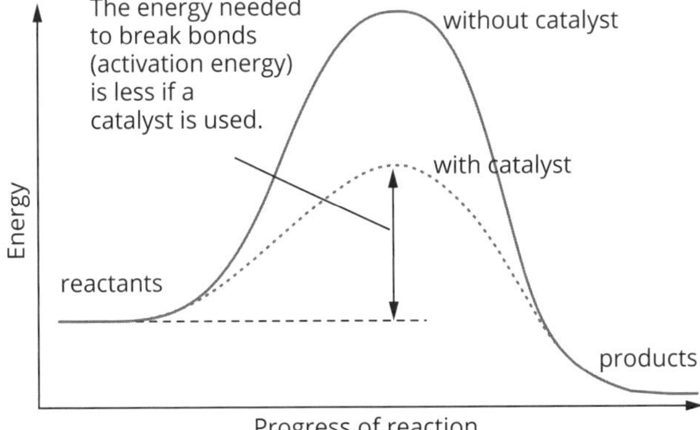

FIGURE 4.6 Energy profile diagrams with and without a catalyst

The shape of a **Maxwell–Boltzmann curve** does not change in the presence of a catalyst (Figure 4.7). However, the lower activation energy means a greater proportion of collisions can overcome the activation energy barrier and so speed up the rate of the reaction.

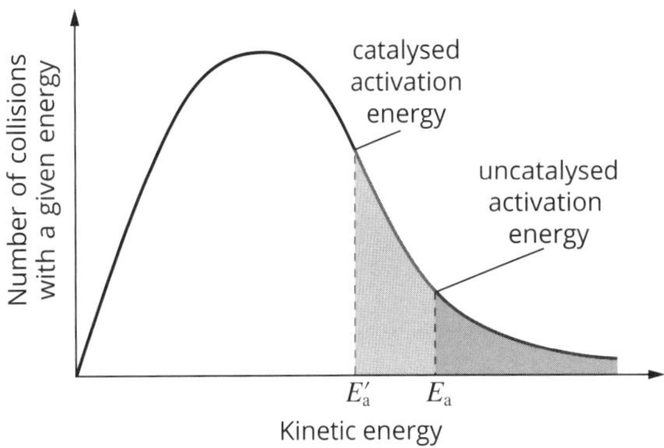

FIGURE 4.7 A catalyst provides a reaction pathway with a lower activation energy, increasing the proportion of collisions that exceed the activation energy and leading to a reaction.

Types of catalysts

Catalysts can be divided into two groups based on the physical state of the chemicals involved:

- Heterogeneous catalysts are in a different physical state from the reactants and products of the reaction.
- Homogeneous catalysts are in the same physical state as the reactants and products of the reaction.

Enthalpy and Hess's law

LATENT HEAT

Latent heat is the energy absorbed or released by a substance as it undergoes a change of state.

Latent heat values are a measure of the enthalpy, or quantity of heat energy, required to melt or boil a given amount of a solid or liquid at its melting or boiling temperature. These enthalpy values are given the symbols ΔH_{vap} and ΔH_{fus} and can have the units kilojoules per kilogram ($kJ\,kg^{-1}$) or kilojoules per mole ($kJ\,mol^{-1}$) (Table 4.2).

TABLE 4.2 Enthalpy values for water

Type of latent heat	Definition	For water ($kJ\,mol^{-1}$)
enthalpy of fusion	heat needed to change a specified quantity of the substance from a solid to a liquid at its melting point	$\Delta H_{fus} = 6.0$
enthalpy of vaporisation	heat needed to change a specified quantity of the substance from a liquid to a gas at its boiling point	$\Delta H_{vap} = 44.0$

The enthalpy values of water are positive and relatively high. This is due to the strength of water's hydrogen bonds. Both fusion and vaporisation are endothermic processes and absorb energy as they proceed.

BOND ENERGY

Bond energy is the amount of energy required to break one mole of bonds in gaseous molecules into their individual atoms under conditions of 25°C and 1 atm of pressure. Bond breaking is always an endothermic process. Bond energy is also the amount of energy released as bonds form. Bond forming is always an exothermic process.

The average bond energies for a selection of single and multiple bonds have been calculated. A selection are listed in Table 4.3.

TABLE 4.3 Average bond energies for some single and multiple bonds

Bond	Bond energy (kJ mol^{-1})
C–H	413
C–C	348
C–Cl	328
H–F	567
H–Cl	431
Cl–Cl	242
C=O	799
C≡C	1072
N≡N	941
O=O	495

The enthalpy change associated with a chemical reaction can be determined by looking at the difference between the energy absorbed to break each bond in reactants and then released as new bonds form in the products.

For example (Table 4.4):

$$CH_4(g) + Cl_2(g) \rightarrow CH_3Cl(g) + HCl(g)$$

TABLE 4.4 Calculating enthalpy change using average bond energies

Average bond energy (kJ mol^{-1})		
bonds broken in reactants	C–H Cl–Cl	+413 +242
	total energy absorbed	413 + 242 = +655 kJ mol^{-1}
bonds formed in products	C–Cl H–Cl	−328 −431
	total energy released	−328 + −431 = −759 kJ mol^{-1}
overall enthalpy change		−759 + 655 = −104 kJ mol^{-1} More energy is released than absorbed so this is an exothermic reaction.
thermochemical equation	$CH_4(g) + Cl_2(g) \rightarrow CH_3Cl(g) + HCl(g)$ $\Delta H = -104$ kJ mol^{-1}	

HESS'S LAW

Hess's law states:

- The amount of heat energy released or absorbed in a chemical reaction is constant irrespective of the number or kind of reaction steps, provided that the same reactants and products are involved.
- The number of steps in the reaction pathway does not affect the overall enthalpy change of the reaction (Figure 4.8).

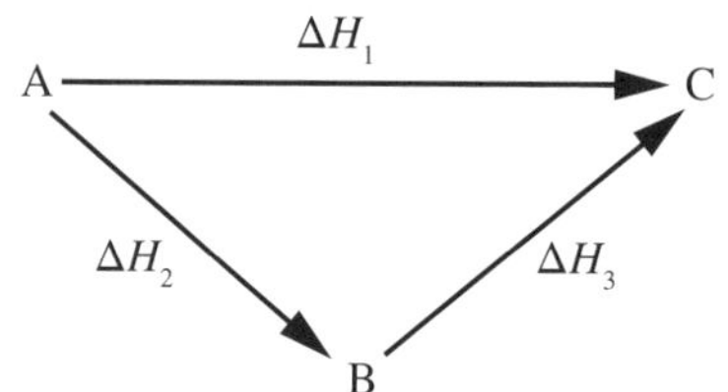

FIGURE 4.8 In this reaction, $\Delta H_1 = \Delta H_2 + \Delta H_3$. The pathway from A to C does not affect the overall enthalpy. This energy cycle diagram demonstrates Hess's law.

For example, methane, CH_4, undergoes complete combustion in oxygen to form carbon dioxide and water according to the equation:

$$CH_4(g) + 2O_2(g) \rightarrow CO_2(g) + 2H_2O(l)$$
$$\Delta H = -890 \text{ kJ mol}^{-1}$$

One reaction pathway can be considered as proceeding via the formation of carbon monoxide:

$$CH_4(g) + 2O_2(g) \rightarrow CO(g) + 2H_2O(l) + \tfrac{1}{2}O_2(g) \quad \Delta H = -606 \text{ kJ mol}^{-1}$$

$$CO(g) + \tfrac{1}{2}O_2(g) \rightarrow CO_2(g) \quad \Delta H = -283 \text{ kJ mol}^{-1}$$

Regardless of the pathway the overall enthalpy change is the same (Figure 4.9).

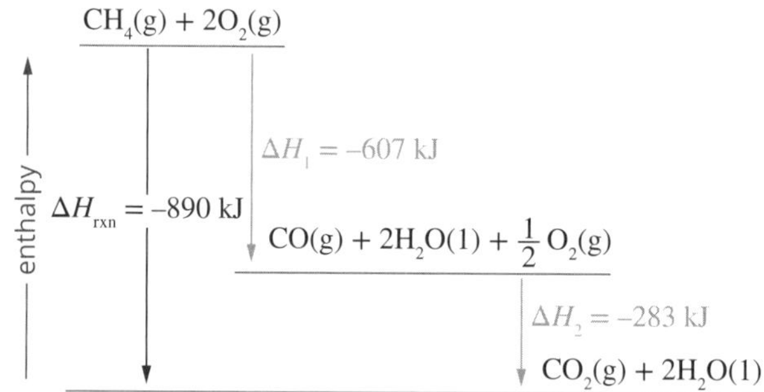

FIGURE 4.9 Whichever reaction pathway is taken to form carbon dioxide and water, the overall change in enthalpy is the same.

The enthalpy changes of a huge range of chemical reactions including oxidation reactions and combustion reactions can be calculated from data about the enthalpy change that occurs for breaking and forming individual bonds. These calculations are based on the following:

- Chemical reactions are reversible if the appropriate conditions are available.
- Energy is always released or absorbed in a chemical reaction. When a reaction is reversed, the magnitude of the enthalpy change is the same; however, the sign of ΔH is changed.
- Equations for chemical reactions can be added and subtracted in the manner of ordinary algebraic equations.

STANDARD ENTHALPY OF FORMATION

The change of enthalpy that occurs when one mole of a compound is formed in its standard state is referred to as **standard enthalpy of formation** and given the symbol ΔH°_f (Figure 4.10).

 ISBN 978 1 4886 1933 5

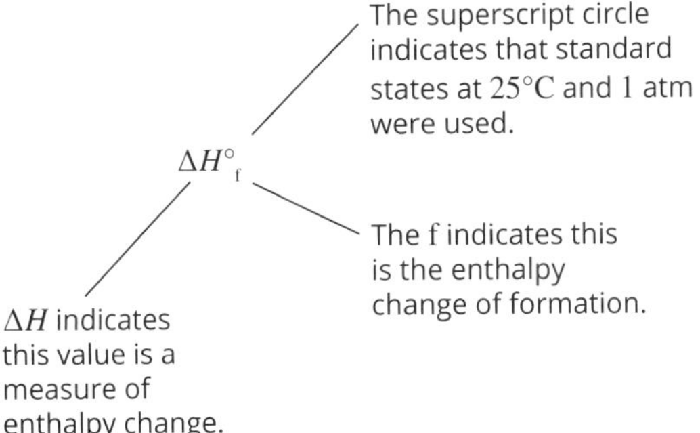

FIGURE 4.10 The symbol for standard enthalpy of formation provides specific information.

Standard enthalpies of formation are a reference that can be used to determine the change in enthalpy for a thermochemical equation (ΔH°) according to the equation:

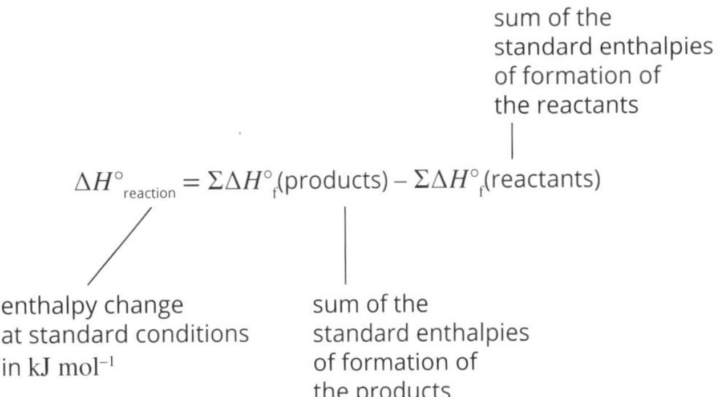

The energy cycles that occur in three different chemical reactions and the calculation of associated enthalpy changes are shown in Table 4.5.

TABLE 4.5 Applications of Hess's law to simple energy cycles

Reaction	Description	Calculation of enthalpy change	Relevant standard enthalpies of formation	Calculations
combustion	occurs when a fuel burns in oxygen releasing energy	$\Delta H^{\circ}_{reaction} = \Sigma\Delta H^{\circ}_{f}(\text{products}) - \Sigma\Delta H^{\circ}_{f}(\text{reactants})$	Complete combustion of methane: $CH_4(g) + 2O_2(g) \rightarrow CO_2(g) + 2H_2O(l)$ $C(s) + O_2(g) \rightarrow CO_2(g)$ $\Delta H^{\circ}_{f} = -393.5$ kJ mol^{-1} $H_2(g) + \frac{1}{2}O_2(g) \rightarrow H_2O(l)$ $\Delta H^{\circ}_{f} = -285.8$ kJ mol^{-1} $C(s) + 2H_2(g) \rightarrow CH_4(g)$ $\Delta H^{\circ}_{f} = -74.8$ kJ mol^{-1} ΔH°_{f} for $O_2 = 0$	$\Sigma\Delta H^{\circ}_{f}(\text{reactants})$ $= 1 \times \Delta H^{\circ}_{f}(CH_4) + 2 \times \Delta H^{\circ}_{f}(O_2)$ $= 1 \times -74.8 + 2 \times 0$ $= -74.8$ kJ mol^{-1} $\Sigma\Delta H^{\circ}_{f}(\text{products})$ $= 1 \times \Delta H^{\circ}_{f}(CO_2) + 2 \times \Delta H^{\circ}_{f}(H_2O)$ $= 1 \times -393.5 + 2 \times -285.8$ $= -965.1$ kJ mol^{-1} $\Delta H^{\circ}_{reaction} = \Sigma\Delta H^{\circ}_{f}(\text{products}) - \Sigma\Delta H^{\circ}_{f}(\text{reactants})$ $= -965.1 - (-74.8)$ $= -890$ kJ mol^{-1}
photosynthesis	process used by plants and some bacteria to use sunlight to make glucose using carbon dioxide and water		Overall equation: $6H_2O(l) + 6CO_2(g) \xrightarrow{\text{sunlight}} C_6H_{12}O_6(s) + 6O_2(g)$ Standard enthalpies of formation: $C(s) + O_2(g) \rightarrow CO_2(g)$ $\Delta H^{\circ}_{f} = -393.5$ kJ mol^{-1} $H_2(g) + O_2(g) \rightarrow H_2O(l)$ $\Delta H^{\circ}_{f} = -285.8$ kJ mol^{-1} $6C(s) + 6H_2(g) + 3O_2(g) \rightarrow C_6H_{12}O_6(s)$ $\Delta H^{\circ}_{f} = -1260$ kJ mol^{-1}	$\Sigma\Delta H^{\circ}_{f}(\text{reactants})$ $= 6 \times \Delta H^{\circ}_{f}(H_2O) + 6 \times \Delta H^{\circ}_{f}(CO_2)$ $= 6 \times -285.8 + 6 \times -393.5$ $= -4075.8$ kJ mol^{-1} $\Sigma\Delta H^{\circ}_{f}(\text{products})$ $= 1 \times \Delta H^{\circ}_{f}(C_6H_{12}O_6) + 6 \times \Delta H^{\circ}_{f}(O_2)$ $= 1 \times -1260 + 6 \times 0$ $= -1260$ kJ mol^{-1} $\Delta H^{\circ}_{reaction} = \Sigma\Delta H^{\circ}_{f}(\text{products}) - \Sigma\Delta H^{\circ}_{f}(\text{reactants})$ $= -1260 - (-4075.8)$ $= +2816$ kJ mol^{-1}
respiration	reaction between glucose and oxygen that takes place in all cells to release energy		Overall equation: $C_6H_{12}O_6(s) + 6O_2(g) \rightarrow 6H_2O(l) + 6CO_2(g)$ Standard enthalpies of formation: $C(s) + O_2(g) \rightarrow CO_2(g)$ $\Delta H^{\circ}_{f} = -393.5$ kJ mol^{-1} $H_2(g) + O_2(g) \rightarrow H_2O(l)$ $\Delta H^{\circ}_{f} = -285.8$ kJ mol^{-1} $6C(s) + 6H_2(g) + 3O_2(g) \rightarrow C_6H_{12}O_6(s)$ $\Delta H^{\circ}_{f} = -1260$ kJ mol^{-1}	$\Sigma\Delta H^{\circ}_{f}(\text{reactants})$ $= 1 \times \Delta H^{\circ}_{f}(C_6H_{12}O_6) + 6 \times \Delta H^{\circ}_{f}(O_2)$ $= 1 \times -1260 + 6 \times 0$ $= -1260$ kJ mol^{-1} $\Sigma\Delta H^{\circ}_{f}(\text{products})$ $= 6 \times \Delta H^{\circ}_{f}(H_2O) + 6 \times \Delta H^{\circ}_{f}(CO_2)$ $= 6 \times -285.8 + 6 \times -393.5$ $= -4075.8$ kJ mol^{-1} $\Delta H^{\circ}_{reaction} = \Sigma\Delta H^{\circ}_{f}(\text{products}) - \Sigma\Delta H^{\circ}_{f}(\text{reactants})$ $= -4075.8 - (-1260)$ $= -2816$ kJ mol^{-1}

Entropy and Gibbs free energy

A **system** includes anything that a scientists wishes to include in a study. The system's **surroundings** is everything outside the boundary of the system. There are three types of system:

- open system—energy and matter can be exchanged with the surroundings
- closed system—only energy and not matter is exchanged with the surroundings
- isolated system—neither matter or energy can be exchanged with the surroundings.

All systems have energy in the form of kinetic energy and internal energy. All systems also have entropy.

ENTROPY

Entropy is a measure of the number of possible arrangements of components available within a system. It measures the order or disorder of a system. A system that is increasing in entropy is said to have increasing disorder. A system with lower entropy is said to have greater order. Entropy is affected by a number of factors:

- A larger system has more entropy than a smaller system as there are more possible arrangements for its components.
- A hotter system has more entropy than a cooler system because the components move more freely and so can have more possible arrangements.
- A liquid has more entropy than the corresponding solid as the particles in a liquid move more freely.
- A gas has more entropy than the corresponding liquid as the gas particles can occupy all of the available space.
- A dissolved solid has more entropy than the solid as the solute particles can be arranged in different ways in the solvent.

Entropy is given the symbol S. The change of entropy in a system (ΔS) can be calculated:

$$\Delta S = S(\text{products}) - S(\text{reactants})$$

A negative ΔS represents a decrease in entropy, i.e. products' entropy < reactants' entropy.

A positive ΔS represents an increase in entropy, i.e. products' entropy > reactants' entropy

A comparison can be made between the concepts of entropy and enthalpy (Table 4.6).

Predicting entropy for a balanced chemical equation

Whether the entropy of a chemical reaction will increase or decrease can be predicted from a balanced chemical equation by considering the following:

- Do the products have more or less particles than the reactants? The side with the greater number of particles has more entropy.
- Do the products have different states to the reactants? The side with more particles in the liquid and gas states has more entropy.

  For example:

$$2H_2(g) + O_2(g) \rightarrow 2H_2O(g)$$

- There are three particles on the reactant side and two on the product side.
- All chemicals present are in the gaseous state so this does not contribute to a change in entropy.
- The reactant side therefore has greater entropy than the product side
- ΔS will be negative.

ENTROPY AND SPONTANEOUS PROCESSES

A spontaneous process is one that tends to occur without the addition of external energy.

- Exothermic reactions tend to be spontaneous.
- Reactions in which there is an increase in entropy tend to be spontaneous.
- Reactions in which there is a decrease in entropy can be spontaneous if the reaction causes a greater entropy increase in the surroundings than the entropy decrease in the system.

The change in entropy of the surroundings is defined as the amount of heat added to the surroundings at a given temperature (T) in kelvin.

As the temperature of any substance is decreased, its particles move more slowly and entropy is decreased. At absolute zero, the particles do not move and are locked in a set position. At this point some disorder remains as there are different possible arrangements of the still particles.

The only way to have zero entropy is for a substance to be at absolute zero with its particles arranged in a perfect network.

TABLE 4.6 Differences between enthalpy and entropy of a chemical reaction

	Enthalpy	Entropy
Definition	difference in energy needed to break bonds in reactants and create bonds in products	measure of the possible arrangements of particles in a system
Symbol	H, ΔH	S, ΔS
Unit	$kJ\,mol^{-1}$	$J\,mol^{-1}\,K^{-1}$
Relevant formula	$\Delta H = H(\text{products}) - H(\text{reactants})$	$\Delta S = S(\text{products}) - S(\text{reactants})$
Charge	$+\Delta H$ represents an endothermic reaction, i.e. the products have more energy than the reactants. $-\Delta H$ represents an exothermic reaction, i.e. the reactants have more energy than the products.	$+\Delta S$ represents a reaction that is becoming more disordered, i.e. the product particles have more possible arrangements than the reactant particles. $-\Delta S$ represents a reaction that is becoming more ordered, i.e. the product particles have less possible arrangements than the reactant particles.

ISBN 978 1 4886 1933 5

Standard entropy values

The absolute entropy of a substance is a value relative to the zero enthalpy of the substance. It is calculated at standard state conditions and given the symbol S°. Some standard entropy values are listed in Table 4.7.

TABLE 4.7 Standard absolute entropies of several substances.

Substance	Formula	S° (J mol^{-1} K^{-1})
carbon dioxide	$CO_2(g)$	+213.8
ethene	$C_2H_4(g)$	+220
oxygen	$O_2(g)$	+205
water	$H_2O(l)$	+70.0

Standard absolute enthalpies can be used to determine the change in entropy for a process according to the equation:

$$\Delta S^\circ = \Sigma S^\circ(\text{products}) - \Sigma S^\circ(\text{reactants})$$

Worked example 3.1: Determine the entropy change for the following reaction.

$$C_2H_4(g) + 3O_2(g) \rightarrow 2CO_2(g) + 2H_2O(l)$$

Thinking	Working
Calculate ΣS° (reactants).	From Table 4.7: $\Sigma S^\circ(\text{reactants}) = 1 \times S^\circ(C_2H_4) + 3 \times S^\circ(O_2)$ $= 1 \times +220 + 3 \times +205$ $= +835\,\text{J mol}^{-}\text{K}^{-1}$
Calculate ΣS° (products).	From Table 4.7: $\Sigma S^\circ(\text{products}) = 2 \times S^\circ(CO_2) + 2 \times S^\circ(H_2O)$ $= 2 \times +213.8 + 2 \times +70.0$ $= +567.6\,\text{J mol}^{-1}\text{K}^{-1}$
Calculate ΔS°.	$\Delta S^\circ = \Sigma S^\circ(\text{products}) - \Sigma S^\circ(\text{reactants})$ $= +567.6 - (835)$ $= -267.4\,\text{J mol}^{-1}\text{K}^{-1}$ The change in entropy for this system at standard state conditions is $-267.4\,\text{J mol}^{-1}\text{K}^{-1}$.

GIBBS FREE ENERGY

Gibbs free energy is a quantity that can be calculated for a particular chemical process using the change in enthalpy (ΔH), temperature (T) of the system and change in entropy (ΔS). Gibbs free energy has the symbol ΔG and the unit kJ mol^{-1}. A useful formula connects the four quantities:

$$\Delta G^\circ = \Delta H^\circ - T\Delta S^\circ$$

The sign of G° can be calculated to determine whether a reaction is spontaneous or non-spontaneous (Table 4.8).

TABLE 4.8 Meaning of different signs of Gibbs free energy

Sign of Gibbs free energy	Prediction of spontaneity
$\Delta G^\circ < 0$	Reaction is spontaneous.
$\Delta G^\circ = 0$	Reaction is neither spontaneous or non-spontaneous.
$\Delta G^\circ > 0$	Reaction is non-spontaneous.

The spontaneity of a chemical reaction is thus dependent on:

- enthalpy change
- entropy change
- temperature.

Effect of temperature on spontaneity of a reaction

When ΔH° and ΔS° have the same sign, the temperature of the reaction determines whether it will be spontaneous. By setting Gibbs free energy in the formula as zero, the minimum temperature at which a reaction will change from being non-spontaneous to spontaneous can be calculated.

Worked example 3.2: At what temperature is the following reaction spontaneous?

$SBr_4(g) \rightarrow S(g) + 2Br_2(l)$

$\Delta H^\circ = +115$ kJ mol^{-1}

$\Delta S^\circ = +125\,\text{J mol}^{-1}\text{K}^{-1} = 0.125\,\text{kJ mol}^{-1}\text{K}^{-1}$

$\Delta G^\circ = 0$

Thinking	Working
Use the formula: $\Delta G^\circ = \Delta H^\circ - T\Delta S^\circ$	$0 = +115 - T \times 0.125$ $-115 = -T \times 0.125$ $T = \frac{115}{0.125} = 920\,\text{K}$ The reaction will become spontaneous at 920 K or 647°C.

The Gibbs free energy does not give any indication of the rate of a chemical reaction.

WORKSHEET 4.1

Knowledge review—predicting products of reactions

Chemistry reactions involve the rearrangement of atoms in reactants to form new products. Predict the products of each of the reactions in the table, give a reason for each prediction, and then write a balanced chemical equation for each reaction.

Reaction number	Reactants	Predicted products	Reasons for prediction
1	zinc and oxygen gas		
2	dilute hydrochloric acid and magnesium metal		
3	solutions of sodium hydroxide and sulfuric acid		
4	propane, C_3H_8, and oxygen gases		
5	calcium carbonate and nitric acid		
6	sodium carbonate and heat		
7	the fuel methanol, CH_3OH, and oxygen gas		
8	solutions of sodium chloride and silver nitrate		

Full balanced equations (don't forget states):

Reaction 1: ______________________________

Reaction 2: ______________________________

Reaction 3: ______________________________

Reaction 4: ______________________________

Reaction 5: ______________________________

Reaction 6: ______________________________

Reaction 7: ______________________________

Reaction 8: ______________________________

 ISBN 978 1 4886 1933 5

WORKSHEET 4.2

Combustion of fuels—thermochemical equations and specific heat capacity

Use the terms in the box to help you complete this worksheet. Some terms may be used more than once and others not at all.

chemical energy	enthalpy	activation energy	energy source	exothermic
endothermic	H	ΔH	$H_{products} - H_{reactants}$	negative
positive	heat capacity	joules	ΔT	

The combustion of fuels releases a significant quantity of heat, which can be used as an ______________________ ______________________. ______________________ ______________________ is the energy stored in the bonds between atoms and between molecules arising from attractions and repulsions between subatomic particles. It is called ______________________ and is represented by the symbol __________. All combustion reactions are ______________________ reactions. In such reactions, the enthalpy change, symbol __________, has a ______________________ sign. ΔH = ______________________.

2 Energy profile diagrams can be used to represent the energy changes in chemical reactions. Draw and label an energy profile diagram for the following reaction using the terms and symbols in the box.

$$CH_4(g) + 2O_2(g) \rightarrow CO_2(g) + 2H_2O(l) \qquad \Delta H = -890\,kJ\,mol^{-1}$$

activation energy	energy	$CH_4(g) + 2O_2(g)$	$CO_2(g) + 2H_2O(l)$
$\Delta H = -890\,kJ\,mol^{-1}$	energy absorbed as bonds break	energy released as bonds form	progress of reaction

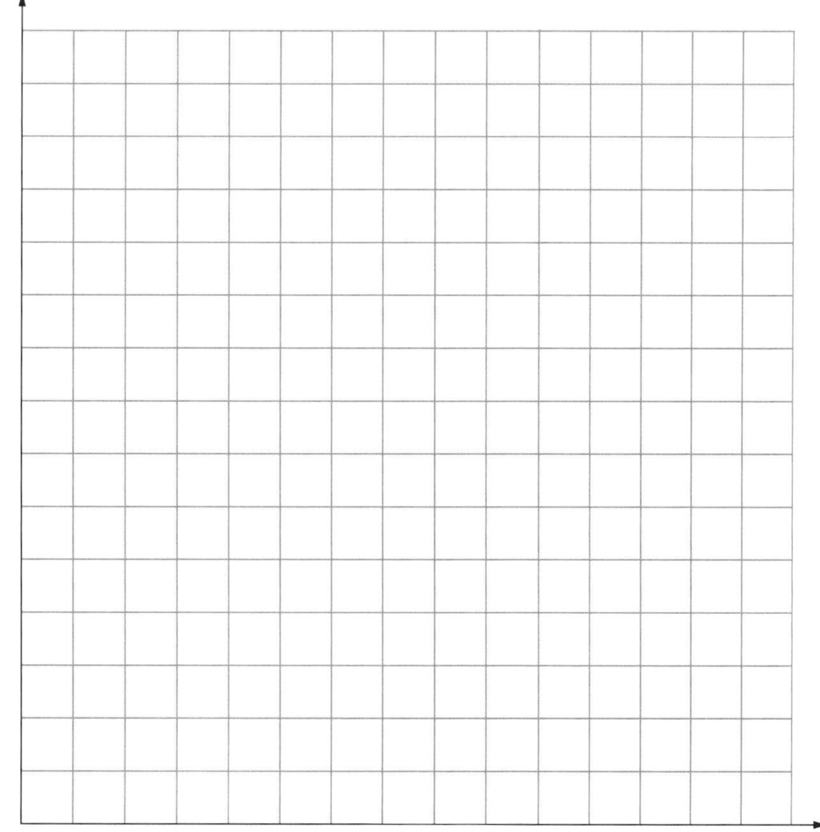

3 Thermochemical equations are balanced chemical equations that include the ΔH for the reaction. Consider the thermochemical equation for octane:

$C_8H_{18}(g) + \frac{25}{2}O_2(g) \rightarrow 8CO_2(g) + 9H_2O(l)$ $\Delta H = -5464\ kJ\ mol^{-1}$

a Determine the value and sign of ΔH for the following equations:

i $2C_8H_{18}(g) + 25O_2(g) \rightarrow 16CO_2(g) + 18H_2O(l)$ $\Delta H =$ ________ $kJ\ mol^{-1}$

ii $4CO_2(g) + \frac{9}{2}H_2O(l) \rightarrow \frac{1}{2}C_8H_{18}(g) + \frac{25}{4}O_2(g)$ $\Delta H =$ ________ $kJ\ mol^{-1}$

b Select the correct alternatives to complete this sentence. The reaction represented by equation i is an *exothermic/endothermic* reaction whereas the reaction represented by equation ii is an *exothermic/endothermic* reaction.

4 Butane gas, C_4H_{10}, can be used as a fuel to heat water. If exactly 1.00 L of water is heated from 20.0°C to 100.0°C, calculate the following using the data in the table. Do your working in the space provided.

a energy absorbed by the water from the combustion of butane

b mass of butane consumed to supply the energy in part **a**.

Specific heat capacity of water	$4.18\ J\ g^{-1}K^{-1}$
Heat of combustion of butane	$-2874\ kJ\ mol^{-1}$
Molar mass of butane	$58.12\ g\ mol^{-1}$
Density of water	$1.00\ g\ mL^{-1}$

Thinking	**Working**
Determine ΔT for the experiment.	$\Delta T =$
a The formula to determine the energy released by the fuel is: energy = specific heat capacity × mass × ΔT	Energy =
b Use the heat of combustion of the fuel to calculate the mass of fuel consumed to produce the energy calculated in part a.	

RATING MY LEARNING	My understanding improved	Not confident ◄──► Very confident ○ ○ ○ ○ ○	I answered questions without help	Not confident ◄──► Very confident ○ ○ ○ ○ ○	I corrected my errors without help	Not confident ◄──► Very confident ○ ○ ○ ○ ○

 ISBN 978 1 4886 1933 5

WORKSHEET 4.3

Investigating enthalpy change—calorimetry

1 The enthalpy change for the combustion of a fuel can be determined using calorimetry. Complete the following statements, which describe steps in this process, using the terms, symbols and formulae from the box. Some terms may be used more than once. Others may not be used at all.

calorimeter	$\frac{E}{\text{amount of fuel (mol)}}$	measured	fuel
$m \times c \times T$	combustion	$\frac{E}{\text{mass of fuel (g)}}$	temperature

i Fill the ______________ with a ______________ volume of water.

ii Weigh the sample of the ______________ and place it in the calorimeter.

ii Record the initial ______________ of the water.

iv Carry out the ______________ of the fuel in the calorimeter.

v Record the final ______________ of the water.

vi Calculate the energy transferred to the water using $E =$ ______________.

vii Calculate the heat of combustion of the fuel in $kJ\,g^{-1}$ using $\Delta H_c =$ ______________.

viii Calculate the heat of combustion of the fuel in $kJ\,mol^{-1}$ using $\Delta H_c =$ ______________.

2 Use the experimental data in the table to calculate the heat of combustion of butane, C_4H_{10}, in $kJ\,mol^{-1}$.

Temperature before combustion	19°C
Temperature after combustion	76°C
Volume of water in calorimeter	100 mL
Mass of butane	0.493 g

RATING MY LEARNING	My understanding improved	Not confident ◄——► Very confident ○ ○ ○ ○ ○	I answered questions without help	Not confident ◄——► Very confident ○ ○ ○ ○ ○	I corrected my errors without help	Not confident ◄——► Very confident ○ ○ ○ ○ ○

WORKSHEET 4.4

Breaking and forming bonds—quantifying enthalpy change

Answer the questions and calculate the enthalpy change associated with the complete combustion of butane gas.

The structure of butane is:

1 Write a balanced chemical equation for the combustion reaction.

2 a Complete the table, including the number of each type of bond in the reactants, the bond energy of each and the number of times each bond type is present in the reactants. Calculate the total energy absorbed by each type of bond to complete the fourth column.

Type of bond	Average bond energy ($kJ\,mol^{-1}$)	Number of times bond occurs in reactants	Enthalpy change ($kJ\,mol^{-1}$)
C–C			
C–H			
O=O			

b Calculate the total enthalpy change when bonds are broken in the reactants.

3 a Complete the table, including the types of bond in the products, the bond energy of each and the number of times each bond type is present in the products. Calculate the total energy needed to form each type of bond in the products to complete the fourth column.

Type of bond	Average bond energy ($kJ\,mol^{-1}$)	Number of times bond occurs in products	Enthalpy change ($kJ\,tmol^{-1}$)
C=O			
O–H			

b Calculate the total enthalpy change when bonds are formed in the products.

4 Calculate the enthalpy change for the reaction and write a thermochemical equation for the reaction.

 ISBN 978 1 4886 1933 5

5 a Is this reaction exothermic or endothermic? ______________________

b Give a reason for your choice.

____

____

6 Sketch an energy profile diagram for this reaction. Label the magnitudes of the activation energy and the ΔH value.

RATING MY LEARNING	My understanding improved	Not confident ⟷ Very confident ○ ○ ○ ○ ○	I answered questions without help	Not confident ⟷ Very confident ○ ○ ○ ○ ○	I corrected my errors without help	Not confident ⟷ Very confident ○ ○ ○ ○ ○

WORKSHEET 4.5

Hess's law—calculating enthalpy changes step by step

1 Explain Hess's law in your own words.

____

____

2 Consider the following reaction:

$C(s) + O_2(g) \rightarrow CO_2(g)$ $\Delta H = -393.5\,kJ\,mol^{-1}$

Calculate the ΔH value for the following reactions:

a $2C(s) + 2O_2(g) \rightarrow 2CO_2(g)$ $\Delta H =$ __________

b $CO_2(g) \rightarrow C(s) + O_2(g)$ $\Delta H =$ __________

c $\frac{1}{2}CO_2(g) \rightarrow \frac{1}{2}C(s) + \frac{1}{2}O_2(g)$ $\Delta H =$ __________

3 Consider the following reaction:

$S(s) + O_2(g) \rightarrow SO_2(g)$ $\Delta H = -296.8\,kJ\,mol^{-1}$

Calculate the ΔH value for the following reactions:

a $2S(s) + 2O_2(g) \rightarrow 2SO_2(g)$ $\Delta H =$ __________

b $SO_2(g) \rightarrow S(s) + O_2(g)$ $\Delta H =$ __________

4 Consider the following reaction:

$C(s) + 2S(s) \rightarrow CS_2(l)$ $\Delta H = +87.9\,kJ\,mol^{-1}$

Calculate the ΔH value for the following reactions:

a $2C(s) + 2S(s) \rightarrow CS_2(l)$ $\Delta H =$ __________

b $CS_2(l) \rightarrow C(s) + 2S(l)$ $\Delta H =$ __________

5 Some combination of the reactions in questions **2–4** occur as steps in the reaction:

$CS_2(l) + 3O_2(g) \rightarrow CO_2(g) + 2SO_2(g)$

a Write out the equations and enthalpies from questions **2–4** that:

i make CO_2 as a product

____

ii make $2SO_2$ as a product

____

iii show CS_2 reacting.

____

b i Combine the three equations from part a into one equation.

____

ii Identify any species that appear as both reactants and products.

____

iii Write the final overall equation.

____

ISBN 978 1 4886 1933 5

c Add the three enthalpies to find the overall enthalpy change.

d Sketch an energy profile diagram for this reaction in the space provided. Label the activation energy and enthalpy change.

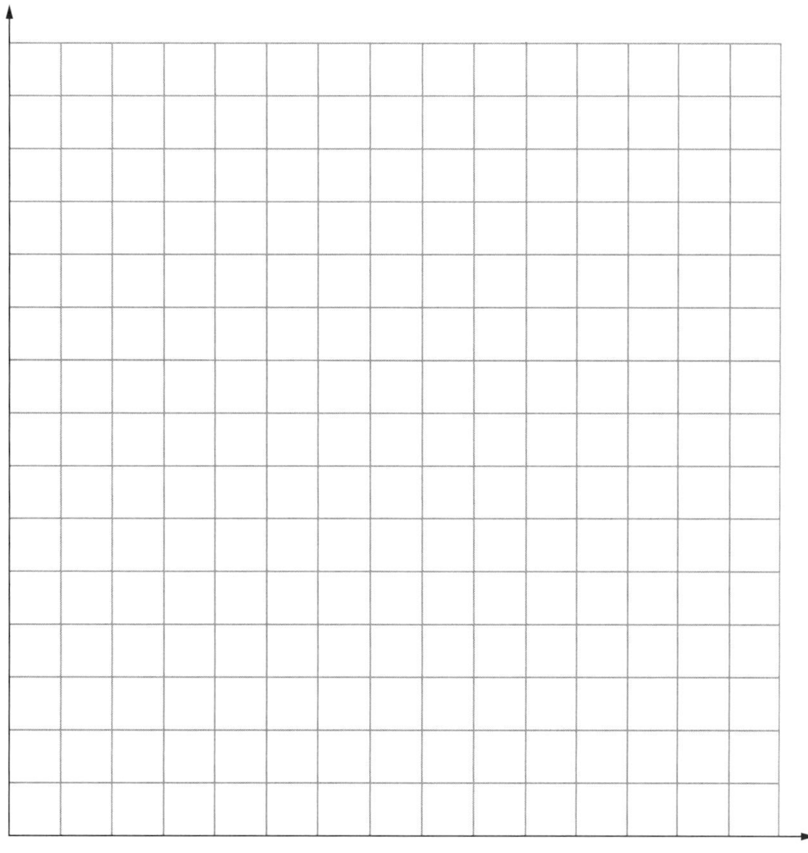

RATING MY LEARNING	My understanding improved	Not confident ◄──► Very confident ○ ○ ○ ○ ○	I answered questions without help	Not confident ◄──► Very confident ○ ○ ○ ○ ○	I corrected my errors without help	Not confident ◄──► Very confident ○ ○ ○ ○ ○

ISBN 978 1 4886 1933 5

WORKSHEET 4.6

Order and disorder—predicting entropy change

For each reaction in the table, predict whether the entropy will increase or decrease as the reaction proceeds. Give a reason for each prediction.

Reaction	Entropy increase or decrease?	Positive or negative ΔS?	Reason for prediction
$(NH_4)_2Cr_2O_7(s) \rightarrow Cr_2O_3(s) + 4H_2O(l) + CO_2(g)$			
$2H_2(g) + O_2(g) \rightarrow 2H_2O(g)$			
$H_2O(s) \rightarrow H_2O(l)$			
$N_2(g) + 3H_2(g) \rightarrow 2NH_3(g)$			
$2Na(s) + 2H_2O(l) \rightarrow 2NaOH(aq) + H_2(g)$			
$H_2O(g) \rightarrow H_2O(l)$			
$MgCO_3(s) \rightarrow MgO(s) + CO_2(g)$			
$NaCl(s) \rightarrow NaCl(aq)$			

RATING MY LEARNING	My understanding improved	Not confident ←→ Very confident ○ ○ ○ ○ ○	I answered questions without help	Not confident ←→ Very confident ○ ○ ○ ○ ○	I corrected my errors without help	Not confident ←→ Very confident ○ ○ ○ ○ ○

ISBN 978 1 4886 1933 5

WORKSHEET 4.7

To be or not to be? Investigating the spontaneity of reactions

1 Classify each of the statements as true or false.

Statement	True or false?
Exothermic reactions are always spontaneous.	
Endothermic reactions are never spontaneous.	
A closed system does not exchange any matter with the surroundings.	
A spontaneous reaction is more likely for a reaction that has an increase in entropy.	
Standard absolute entropy values are determined at 1 atm and 0°C.	
A reaction with a positive entropy change will not necessarily be spontaneous.	
Gibbs free energy is dependent on change in entropy, change in enthalpy and the temperature.	
Increasing entropy is associated with increasing disorder in a system.	
A positive value for Gibbs free energy allows the prediction that a reaction will not be spontaneous.	

2 **a** Write the expression used for the calculation of Gibbs free energy.

____

____

b What are the three factors that contribute to Gibbs free energy?

____

____

____

3 Based on the Gibbs free energy formula, predict the spontaneity of a reaction for each of the following combinations of signs for ΔH and ΔS. Comment on the dependence of the spontaneity on temperature.

ΔH	ΔS	Will reaction be spontaneous? How dependent is spontaneity on temperature?
negative	negative	
negative	positive	
positive	negative	
positive	positive	

RATING MY LEARNING	My understanding improved	Not confident ◄──► Very confident ○ ○ ○ ○ ○	I answered questions without help	Not confident ◄──► Very confident ○ ○ ○ ○ ○	I corrected my errors without help	Not confident ◄──► Very confident ○ ○ ○ ○ ○

WORKSHEET 4.8

Literacy review—comparing key terms

Compare the concepts represented by the following pairs of key terms:

1 exothermic and endothermic

2 chemical equation and thermochemical equation

3 bond forming and bond breaking

4 enthalpy change and entropy change

5 $kJ\,mol^{-1}$ and $J\,mol^{-1}\,K^{-1}$

6 $-\Delta H$ and $+\Delta H$

7 $-\Delta S$ and $+\Delta S$

8 $-\Delta G$ and $+\Delta G$

RATING MY LEARNING	My understanding improved	Not confident ◄—► Very confident ○ ○ ○ ○ ○	I answered questions without help	Not confident ◄—► Very confident ○ ○ ○ ○ ○	I corrected my errors without help	Not confident ◄—► Very confident ○ ○ ○ ○ ○

ISBN 978 1 4886 1933 5

WORKSHEET 4.9

Thinking about my learning

On completion of Module 4: Drivers of reactions, you should be able to describe, explain and apply the relevant scientific ideas. You should be able to interpret, analyse and evaluate data.

1 The table shows the areas of key knowledge covered in this module. Reflect on how well you understand each area. Rate your learning by shading the circle that corresponds to your level of understanding for each section. It may be helpful to use colour as a visual representation. For example:

- green—very confident
- orange—in the middle
- red—starting to develop.

Section focus	**Rate my learning**				
	Starting to develop ◄				► Very confident
Energy changes in chemical reactions—enthalpy, exothermic and endothermic reactions, energy profile diagrams, activation energy, thermochemical equations	○	○	○	○	○
Heat of combustion—$kJ\ mol^{-1}$ and $kJ\ g^{-1}$	○	○	○	○	○
Calorimetry—bomb and solution, enthalpy of dissolution	○	○	○	○	○
Catalysts—effect on activation energy, types	○	○	○	○	○
Hess's law—bond energy, reaction pathways, steps in a reaction pathway, standard enthalpy of formation	○	○	○	○	○
Entropy—system and surroundings, causes of increasing entropy, comparison with enthalpy	○	○	○	○	○
Spontaneous and non-spontaneous processes—Gibbs free energy, effect of temperature	○	○	○	○	○

2 Consider the points you have shaded from 'starting to develop' to 'in the middle'. List specific ideas you can identify that were challenging.

____

____

____

____

____

____

3 Write down two different strategies that you will apply to help further your understanding of these ideas.

____

____

____

____

____

PRACTICAL ACTIVITY 4.1

Exothermic and endothermic reactions

Suggested duration: 30 minutes

INTRODUCTION

Reactions that release energy are exothermic and reactions that absorb energy are endothermic. Measuring temperature change over time for a reaction, using a thermometer or temperature probe, can determine whether the reaction is exothermic or endothermic.

PURPOSE

To observe the temperature changes associated with chemical reactions.

PRE-LAB SAFETY INFORMATION

Material used	Hazard	Control
1 mol L^{-1} hydrochloric acid	toxic by all routes of exposure; lung irritant	Wear safety glasses, gloves and a laboratory coat.
1 mol L^{-1} sodium hydroxide solution	corrosive	Wear safety glasses, gloves and a laboratory coat.
1.5 mol L^{-1} citric acid	serious eye and respiratory irritant	Wear safety glasses, gloves and a laboratory coat.
1 mol L^{-1} copper (II) sulfate	eye and skin irritant; toxic to environment	Wear safety glasses, gloves and a laboratory coat. Use appropriate disposal method.
copper(II) sulfate	toxic; serious skin and eye irritant; toxic to environment	Wear safety glasses, gloves and a laboratory coat. Use appropriate disposal method.
sodium hydrogen carbonate	slightly toxic	Wear safety glasses, gloves and a laboratory coat.
magnesium turnings	contact with water releases flammable gases	Wear safety glasses, gloves and a laboratory coat. Do not use near open flame, sparks, etc.
potassium nitrate	serious eye, skin and lung irritant; oxidiser; may intensify fire; forms explosive mixtures with active metals	Wear safety glasses, gloves and a laboratory coat. Restrict use to designated procedure.
sodium thiosulfate	slightly toxic	Wear safety glasses, gloves and a laboratory coat.

Please indicate that you have understood the information in the safety table.

Name (print): ______________________

I understand the safety information (signature): ______________________

MATERIALS

- 10 mL 1 mol L^{-1} copper(II) sulfate solution, $CuSO_4$
- 10 mL 1.5 mol L^{-1} citric acid solution, $C_6H_8O_7$
- 5 mL 1 mol L^{-1} hydrochloric acid, HCl
- 5 mL 1 mol L^{-1} sodium hydroxide solution, NaOH
- solid sodium hydrogen carbonate, $NaHCO_3$
- solid anhydrous copper(II) sulfate, $CuSO_4$
- magnesium turnings
- solid potassium nitrate, KNO_3
- solid sodium thiosulfate, $Na_2S_2O_3$
- 6 test-tubes
- test-tube rack
- thermometer, −10°C to 110°C or temperature probe and data collection system
- 10 mL measuring cylinder
- spatula
- safety glasses

PROCEDURE

1 Pour 5 mL of 1 mol L^{-1} NaOH solution into a test-tube. Record its temperature.

2 Add 5 mL of 1 mol L^{-1} HCl to the test-tube. Use the thermometer to gently stir the liquid. Record the change in temperature that occurs.

ISBN 978 1 4886 1933 5

3 Repeat steps 1 and 2 using the following pairs of reactants (do not investigate any other combination of chemicals; mix only the pairs of chemicals listed):

- 10 mL 1.5 mol L^{-1} citric acid solution and half a spatula of $NaHCO_3$
- 10 mL water and half a spatula of KNO_3
- 10 mL water and half a spatula of anhydrous $CuSO_4$
- 10 mL 1 mol L^{-1} $CuSO_4$ solution and half a spatula of magnesium turnings
- 10 mL water and half a spatula of $Na_2S_2O_3$.

Use of electronic data collection equipment

A temperature probe can be used instead of a thermometer. For the reaction involved in steps 1 and 2, collect data for 5 minutes. Data collection should begin about 20 seconds before the reactants are mixed in step 2. Display the data as a temperature versus time graph. Repeat for the first pair of chemicals in step 3 (and other pairs if time permits).

RESULTS

Construct a table to record your results.

ANALYSIS OF RESULTS

1 Which reactions are endothermic and which are exothermic?

2 For each process indicate whether the enthalpy change (ΔH) is positive or negative.

ISBN 978 1 4886 1933 5

3 Suppose you performed the third reaction in step 3 with 10 mL water and a spatula of anhydrous copper(II) sulfate. Predict the temperature change you would observe.

4 Predict the temperature changes that would occur if you were able to reverse the reactions in this experiment.

DISCUSSION

5 Sketch two energy profile diagrams, one for one of the exothermic processes and one for one of the endothermic processes.

CONCLUSION

RATING MY LEARNING	My understanding improved	Not confident ◄──► Very confident ○ ○ ○ ○ ○	I answered questions without help	Not confident ◄──► Very confident ○ ○ ○ ○ ○	I corrected my errors without help	Not confident ◄──► Very confident ○ ○ ○ ○ ○

 ISBN 978 1 4886 1933 5

PRACTICAL ACTIVITY 4.2

Energy from different fuels

Suggested duration: 30 minutes

INTRODUCTION

Alcohols are useful fuels. In this experiment, the energy released by the combustion of ethanol is used to heat water. Knowing that 4.18 J of energy is required to heat 1 g (1 mL) of water by 1 K, the quantity of energy released by combustion of ethanol can be calculated.

$4.18\,J\,g^{-1}K^{-1}$ is the specific heat capacity of water.

PURPOSE

To measure and compare the thermal energy released during the combustion of ethanol.

PRE-LAB SAFETY INFORMATION

Material used	Hazard	Control
ethanol	toxic and highly flammable	Avoid skin contact. Do not breathe vapour.
filling spirit burners	flammable liquids	Do not fill near open flame.

Please indicate that you have understood the information in the safety table.

Name (print): ______________________

I understand the safety information (signature): ______________________

MATERIALS

- spirit burner containing an alcohol, ethanol (CH_3CH_2OH)
- other alcohols: methanol (CH_3OH), propanol (C_3H_7OH)
- 250 mL measuring cylinder
- steel can (or conical flask)
- thermometer, −10 to 110°C
- retort stand, bosshead and clamp
- bench mat
- matches
- electronic balance
- safety glasses

PROCEDURE

Record all of your procedure results in Table 1.

1. Using a retort stand and clamp, set up the steel can 3–4 cm above the wick of the spirit burner.
2. Pour 200 mL water into the can and record its temperature.
3. Weigh the spirit burner and ethanol. Record the mass of ethanol.
4. Light the burner and heat the water. Stir continuously.
5. When the water temperature has increased by about 20°C, extinguish the burner and record the highest temperature reached by the water.
6. Record the mass of the burner and hence deduce the mass of ethanol consumed. Record this mass.
7. Repeat the experiment using another alcohol.

Use of electronic data equipment

A temperature probe can be used instead of a thermometer. Collect data for 10 minutes. Data collection should begin about 45 seconds before the burner is lit. Extinguish the burner when the temperature rises by about 20°C or after 7 minutes of data collection, whichever occurs first. Display each set of data as a graph of temperature versus time.

ISBN 978 1 4886 1933 5

PRACTICAL ACTIVITY 4.2

RESULTS

TABLE 1 Mass and temperature results

Initial mass of spirit burner and alcohol (g)	
Final mass of spirit burner and alcohol (g)	
Mass of ethanol (g)	
Initial temperature (°C)	
Final temperature (°C)	
ΔT **(°C)**	

ANALYSIS OF RESULTS

1 Write a balanced equation for the combustion of ethanol.

2 Calculate the energy, in kJ, absorbed by the water in the can when the ethanol underwent combustion.

3 Calculate the energy released per gram of ethanol burnt.

4 Calculate the heat of combustion of ethanol in $kJ\,mol^{-1}$.

ISBN 978 1 4886 1933 5

PRACTICAL ACTIVITY 4.2

5 The heat of combustion of ethanol is $-1364\ kJ\ mol^{-1}$.

a How well do your results agree with these values?

b Comment on the main sources of error in this experiment and suggest how the experiment could be improved.

6 Comment on the reliability of conclusions you can draw from this investigation.

CONCLUSION

RATING MY LEARNING	My understanding improved	Not confident ○ ○ ○ ○ ○ Very confident	I answered questions without help	Not confident 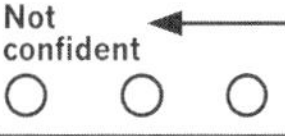 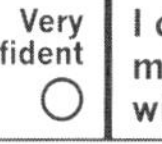 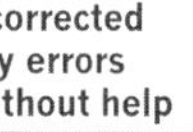 ○ ○ ○ ○ ○ Very confident	I corrected my errors without help	Not confident ○ ○ ○ ○ ○ Very confident

PRACTICAL ACTIVITY 4.3

Enthalpy changes in chemical reactions

Suggested duration: 45 minutes

INTRODUCTION

Temperature changes during a chemical reaction are not the same as the energy released or absorbed. Enthalpy change during a reaction can be measured using a calorimeter.

$4.18\,J\,g^{-1}K^{-1}$ is the specific heat capacity of water.

PURPOSE

To determine energy changes in various chemical reactions using a calibrated calorimeter.

PRE-LAB SAFETY INFORMATION		
Material used	**Hazard**	**Control**
1 mol L^{-1} hydrochloric acid	toxic by all routes of exposure; lung irritant	Wear safety glasses, gloves and a laboratory coat.
1 mol L^{-1} sodium hydroxide solution	corrosive	Wear safety glasses and a laboratory coat.
potassium nitrate	serious eye, skin and lung irritant; oxidiser; may intensify fire; forms explosive mixtures with active metals	Wear safety glasses, gloves and a laboratory coat. Restrict use to designated procedure.
magnesium powder	contact with water releases flammable gases	Wear safety glasses, gloves and a laboratory coat. Do not use near open flame, sparks, etc.

Please indicate that you have understood the information in the safety table.

Name (print): ______________________________

I understand the safety information (signature): ______________________________

MATERIALS

- 3 g solid potassium nitrate (KNO_3), coarsely ground
- 50 mL 1.0 mol L^{-1} sodium hydroxide solution, NaOH
- 150 mL 1.0 mol L^{-1} hydrochloric acid, HCl
- 0.25 g magnesium powder
- 2 × 100 mL measuring cylinders
- calorimeter, previously calibrated if possible
- thermometer, –10°C to 50°C
- spatula
- 2 × weighing bottles or 2 × watchglasses
- electronic balance
- safety glasses

PROCEDURE

Part A—Reaction between magnesium and hydrochloric acid

1 Pour 100 mL of 1.0 mol L^{-1} HCl into a calorimeter. Stir and record the temperature of the acid.

2 Add between 0.20 g and 0.25 g of accurately weighed magnesium powder to the calorimeter. Stir the mixture and record the highest temperature it reaches.

Part B—Reaction between hydrochloric acid and sodium hydroxide solution

1 Pour 50 mL of 1.0 mol L^{-1} NaOH solution into a calorimeter. Stir, and record the temperature of the solution.

2 Pour 50 mL of 1.0 mol L^{-1} HCl into a measuring cylinder and record its temperature. Add the HCl to the NaOH in the calorimeter. Stir the mixture and record the highest temperature it reaches.

Part C—Dissolution of potassium nitrate

1 Pour 100 mL water into a calorimeter. Stir, and record the temperature of the water.

2 Add about 3 g of accurately weighed KNO_3 to the calorimeter. Stir and record the lowest steady temperature reached by the contents of the calorimeter.

ISBN 978 1 4886 1933 5

PRACTICAL ACTIVITY 4.3

Use of electronic data collection equipment

A temperature probe can be used instead of a thermometer. For each of the reactions involved in parts A–C, collect data for 5 minutes. Data collection should begin about 20 seconds before the reactants are mixed in step 2. Display each set of data as a temperature against time graph.

ANALYSIS OF RESULTS

1 Write an equation for each of the reactions that occur in this experiment.

2 Using your measurements of the changes in temperature and the calibration factor of the calorimeter, calculate the energy change, in joules, that occurred during each reaction. (If the calibration factor is unknown, an estimate of the energy change can still be calculated using the fact that 1 g of water requires 4.18 J to raise the temperature by 1 K.)

3 Calculate the heat of reaction (ΔH) for each equation you wrote in your answer to question **1**.

4 For each reaction that occurred in this investigation, which is the greater: the energy required to break the bonds in the reactants or the energy released when new bonds are formed to make the products? To explain your answer, draw a diagram to illustrate the energy change during each reaction.

DISCUSSION

5 Energy is neither created nor destroyed in a chemical reaction. Explain where the energy released (or absorbed) by the reactions comes from (or goes to).

CONCLUSION

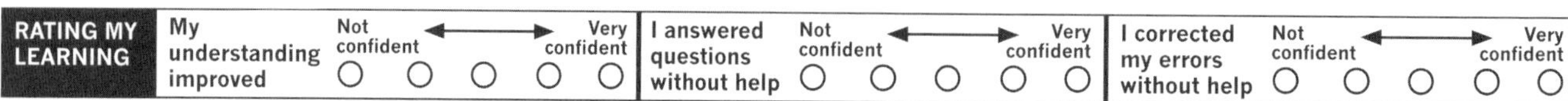

ISBN 978 1 4886 1933 5

PRACTICAL ACTIVITY 4.4

Investigating temperature change, enthalpy and entropy

Suggested duration: 10 minutes

INTRODUCTION

A spontaneous process is one that tends to occur without the addition of external energy. The spontaneity of a chemical reaction is dependent on:

- enthalpy change (exothermic reactions tend to be spontaneous)
- entropy change (reactions in which there is an increase in entropy tend to be spontaneous)
- temperature.

Observations of the enthalpy and entropy changes of a chemical reaction allow conclusions to be drawn about the spontaneity of the reaction.

PURPOSE

To demonstrate exothermic and endothermic reactions, observe the spontaneity of the reactions and consider the effect of entropy.

MATERIALS

- 30 g calcium oxide (CaO), lumps or powder
- 250 mL beaker
- 10 mL measuring cylinder
- bench mat
- thermometer, 0–200°C, preferably a digital thermometer, which allows the rising temperature of the contents of the flask to be more easily monitored
- safety glasses

Part A—Chemical oven

PRE-LAB SAFETY INFORMATION

Material used	Hazard	Control
calcium oxide	corrosive; eye, skin and lung irritant	Wear safety glasses, gloves and a laboratory coat.
calcium hydroxide	corrosive; eye and skin irritant	Wear safety glasses, gloves and a laboratory coat.
hot flask	burns from the hot flask or steam that is produced; if a crust forms in the beaker, trapped steam may cause the contents of the beaker to be suddenly ejected	Do not use more water than specified.

Please indicate that you have understood the information in the safety table.

Name (print): ______________________________

I understand the safety information (signature): ______________________________

PROCEDURE

1 Place the beaker on a bench mat. Place 30 g CaO in the beaker.

2 Add about 8 mL water and insert a thermometer into the mixture.

3 After a few minutes, so much heat is produced that steam should be seen rising from the flask. Note the temperature.

RESULTS

Record the highest temperature reached by the reaction __________ °C.

ANALYSIS OF RESULTS

1 What energy transformation has occurred in the course of the reaction? Where has the energy come from or gone to?

____

____

2 Is the reaction exothermic or endothermic?

____

3 Write an equation for the reaction.

____

4 Would you predict this reaction to have an increase or decrease in entropy as it proceeds? Why?

____

____

____

5 Is the reaction spontaneous or non-spontaneous? What does this tell you about Gibbs free energy for this reaction?

____

____

Part B—Reaction between two solids

THEORY

The equation for this reaction is:

$Ba(OH)_2 \cdot 8H_2O(s) + 2NH_4Cl(s) \rightarrow BaCl_2(aq) + 2NH_3(aq) + 10H_2O(l)$

Although the ΔH for this reaction is positive, the reaction is spontaneous due to an increase in entropy, $\Delta G = \Delta H - T\Delta S$.

PRE-LAB SAFETY INFORMATION

Material used	Hazard	Control
barium hydroxide	poisonous if ingested; skin irritant	Wear safety glasses, gloves and a laboratory coat.
ammonium salts	toxic; serious eye and skin irritant	Wear safety glasses, gloves and a laboratory coat.
ammonia vapour (product)	eye, skin and lung irritant	Take care not to inhale the vapour. Work in a well-ventilated space.

Please indicate that you have understood the information in the safety table.

Name (print): ____________________

I understand the safety information (signature): ____________________

MATERIALS

- 32 g barium hydroxide ($Ba(OH)_2 \cdot 8H_2O$)
- 11 g ammonium chloride, NH_4Cl (or 16 g ammonium thiocyanate)
- 250 mL beaker
- glass stirring rod
- thermometer, −40°C to 50°C, preferably a digital thermometer, which allows the temperature of the contents of the flask to be more easily monitored
- small block of wood or plastic
- woollen gloves
- safety glasses

PROCEDURE

1 Place the pre-weighed solids into the beaker and stir to mix them. Within 30 seconds, the odour of ammonia can be detected. A noticeable amount of frosting occurs on the outside of the beaker as the temperature drops dramatically.

2 Wearing woollen gloves, demonstrate the low temperature of the beaker by wetting a small wooden block and placing the flask on it. The flask quickly freezes to the block.

RESULTS

Record the lowest temperature reached by the reaction: __________ °C.

 ISBN 978 1 4886 1933 5

ANALYSIS OF RESULTS

1 Is the reaction exothermic or endothermic? What does this tell you about the sign of the ΔH value?

2 How does the chemical energy of the reactants compare with the energy of the products?

3 What has happened to some of the thermal energy present in the beaker during the course of the reaction?

4 Is the entropy change, ΔS, for this reaction positive or negative? Explain your answer.

5 In terms of ΔS and ΔG and the relationship $\Delta G = \Delta H - T\Delta S$, explain why this endothermic reaction is spontaneous at all temperatures.

CONCLUSION

RATING MY LEARNING	My understanding improved	Not confident ◄——► Very confident ○ ○ ○ ○ ○	I answered questions without help	Not confident ◄——► Very confident ○ ○ ○ ○ ○	I corrected my errors without help	Not confident ◄——► Very confident ○ ○ ○ ○ ○

ISBN 978 1 4886 1933 5

DEPTH STUDY 4.1

Research investigation into an aspect of enthalpy or entropy

Suggested duration: 3.5 hours

INTRODUCTION

This depth study allows you to choose a question so that you can plan a research investigation on an aspect of drivers of chemical reactions. You will analyse the information, then communicate your findings in an appropriate manner.

QUESTIONING AND PREDICTING

1 Choose one inquiry question from the following list or develop your own question in consultation with your teacher. Highlight your chosen question.

- What are the factors that determine bond energy?
- Who was Germain Henri Hess and how did he develop Hess's law?
- Who was Josiah Willard Gibbs and how was the concept of Gibbs free energy developed?
- What patterns exist in the sizes of different bond energies?
- What are some real-world examples to which the size of a bond energy is particularly relevant?
- What are the most stable molecules?
- Why is energy required to break a chemical bond and released when a bond is formed?
- Who came up with the laws of thermodynamics?

2 Consider your chosen question. What will you need to find out to shape your response? What further questions arise that you will need to research?

ISBN 978 1 4886 1933 5

CONDUCTING YOUR INVESTIGATION

For each source that you access during this research investigation, you should fill in a summary table such as this one. Make as many copies of this table as you need.

Bibliographic information	
Summary of content (be concise and coherent)	
Relevant findings and evidence	
Limitations, bias or flaws within the article	
Useful quotations	
Additional notes	

ANALYSING DATA AND INFORMATION

After you have completed your research summaries and are satisfied you have collected sufficient information, use the following space to plan out the presentation of your results. You should consider questions such as:

- What background concepts need to be included?
- Which chemical terms need to be defined?
- What will you include in the introduction?
- What subheadings will you need to use?
- What images, graphs, tables are relevant and helpful and should be included?
- How will you construct an evidence-based, fully justified answer to your inquiry question?
- Is there an area of future research you could suggest?

COMMUNICATING

Your teacher may ask you to present your research findings in a particular way, or you could choose an appropriate format. For example:

- poster
- oral/PowerPoint presentation
- written report
- webpage.

In order to choose the best format, you should consider your audience.

Whichever format you choose, remember to use suitable language and terminology.

Your final report should contain no more than 1000 words.

 ISBN 978 1 4886 1933 5

MODULE 4 • REVIEW QUESTIONS

Multiple choice

1 Consider the following reaction:

$2H_2(g) + O_2(g) \rightarrow 2H_2O(l) \qquad \Delta H = -572\ kJ\,mol^{-1}$

Which of the following will be the enthalpy for the equation shown?

$H_2O(l) \rightarrow H_2(g) + \frac{1}{2}O_2(g)$

A +527 kJ mol^{-1}

B −236 kJ mol^{-1}

C +236 kJ mol^{-1}

D −1144 kJ mol^{-1}

2 Propane burns in oxygen according to the equation:

$C_3H_8(g) + 5O_2(g) \rightarrow 3CO_2(g) + 4H_2O(l)$

$\Delta H = -2220\ kJ\,mol^{-1}$

What is the mass of propane that must undergo combustion in oxygen in order to heat 200 mL of water from 16°C to 100°C?

A 0.032 g

B 1.39 g

C 31.6 g

D 70.3 g

3 The average bond energies of several bonds are shown in the table.

Bond	Bond energy (kJ mol^{-1})
C–H	413
C–C	348
C=C	614

What amount of energy is needed to break the bonds in the molecule shown?

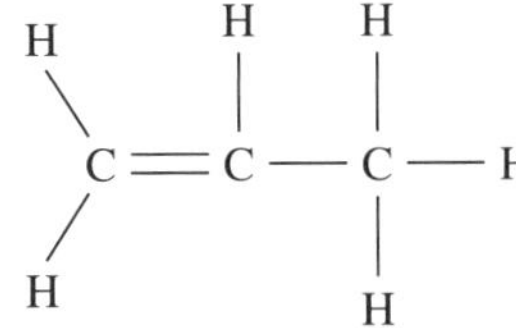

A 614 kJ mol^{-1}

B 1375 kJ mol^{-1}

C 2478 kJ mol^{-1}

D 3440 kJ mol^{-1}

4 Consider the following thermochemical equations:

$C + O_2 \rightarrow CO_2 \qquad \Delta H = -393\ kJ\,mol^{-1}$

$H_2 + \frac{1}{2}O_2 \rightarrow H_2O \qquad \Delta H = -286\ kJ\,mol^{-1}$

$CH_4 + 2O_2 \rightarrow CO_2 + 2H_2O \qquad \Delta H = -892\ kJ\,mol^{-1}$

According to Hess's law, determine the enthalpy for the reaction shown.

$C + 2H_2 \rightarrow CH_4$

A −73 kJ mol^{-1}

B +73 kJ mol^{-1}

C −213 kJ mol^{-1}

D +1857 kJ mol^{-1}

5 Which of the following reactions is most likely to have the greatest increase in entropy?

A $H_2O(s) \rightarrow H_2O(l)$

B $2H_2(g) + O_2(g) \rightarrow 2H_2O(g)$

C $2K(s) + 2H_2O(l) \rightarrow 2KOH(aq) + H_2(g)$

D $N_2(g) + 3H_2(g) \rightarrow 2NH_3(g)$

6 Which combination of enthalpy and entropy values is likely to be a spontaneous reaction at all temperatures?

A $-\Delta H, +\Delta S$

B $-\Delta H, -\Delta S$

C $+\Delta H, +\Delta S$

D $+\Delta H, -\Delta S$

Short answer

7 Reactant A reacts with reactant B to produce product C according to the following equation:

$A(g) + 3B(g) \rightarrow 2C(g) \qquad \Delta H = +48\ kJ\,mol^{-1}$

The activation energy for the forward reaction is 150 kJ mol^{-1}.

a On the axes shown, draw an energy profile diagram for this reaction.

b Add a second line to the energy profile diagram showing the effect of a catalyst.

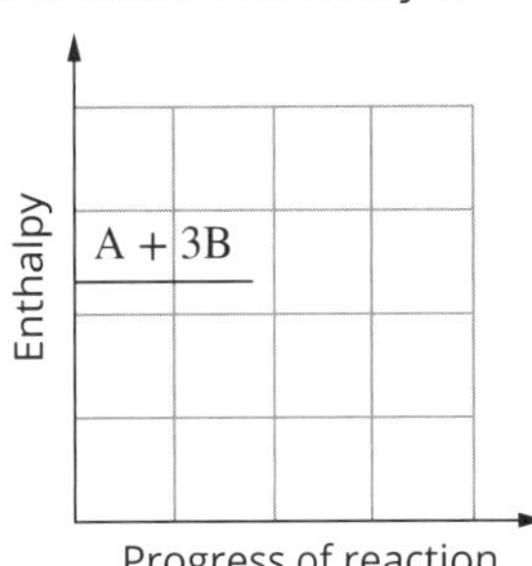

c Do you predict the ΔS for this reaction to be positive or negative? Explain your answer.

ISBN 978 1 4886 1933 5

8 The following reaction is non-spontaneous at 22°C:

$CaCO_3(s) \rightarrow CaO(s) + CO_2(g) \quad \Delta H = +178.49\,kJ\,mol^{-1}$

The entropy change for this reaction is $+165.18\,kJ\,mol^{-1}$.

a Calculate Gibbs free energy for the reaction at 22°C.

b Calculate the temperature, in °C, at which the reaction will become spontaneous.

Extended response

9 Compare the concepts of entropy and enthalpy in terms of their definition, symbol and unit. Also compare the meaning of the sign when a change in each is measured.

 ISBN 978 1 4886 1933 5

10 A student investigates the energy released by the combustion of different fuels. She sets up the apparatus as shown.

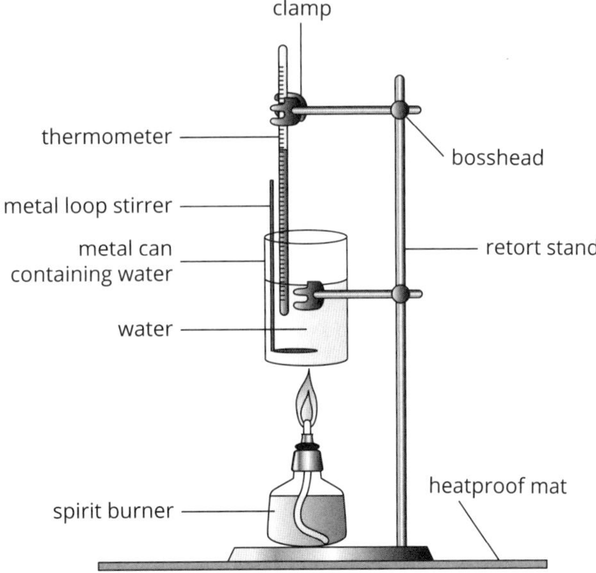

First she burns a 5 mL sample of ethanol, which heats half a can of water by 7.56°C. She discards that water and refills the can from the tap. She then burns a 5 mL sample of methanol, which heats the water by 5.65°C.

The student reaches the conclusion that ethanol has a higher heat of combustion, in $kJ\,g^{-1}$, than methanol.

a Is the conclusion reached by the student valid? Explain your answer.

b Describe three ways you would improve the design of this practical investigation. Give a reason for each improvement.

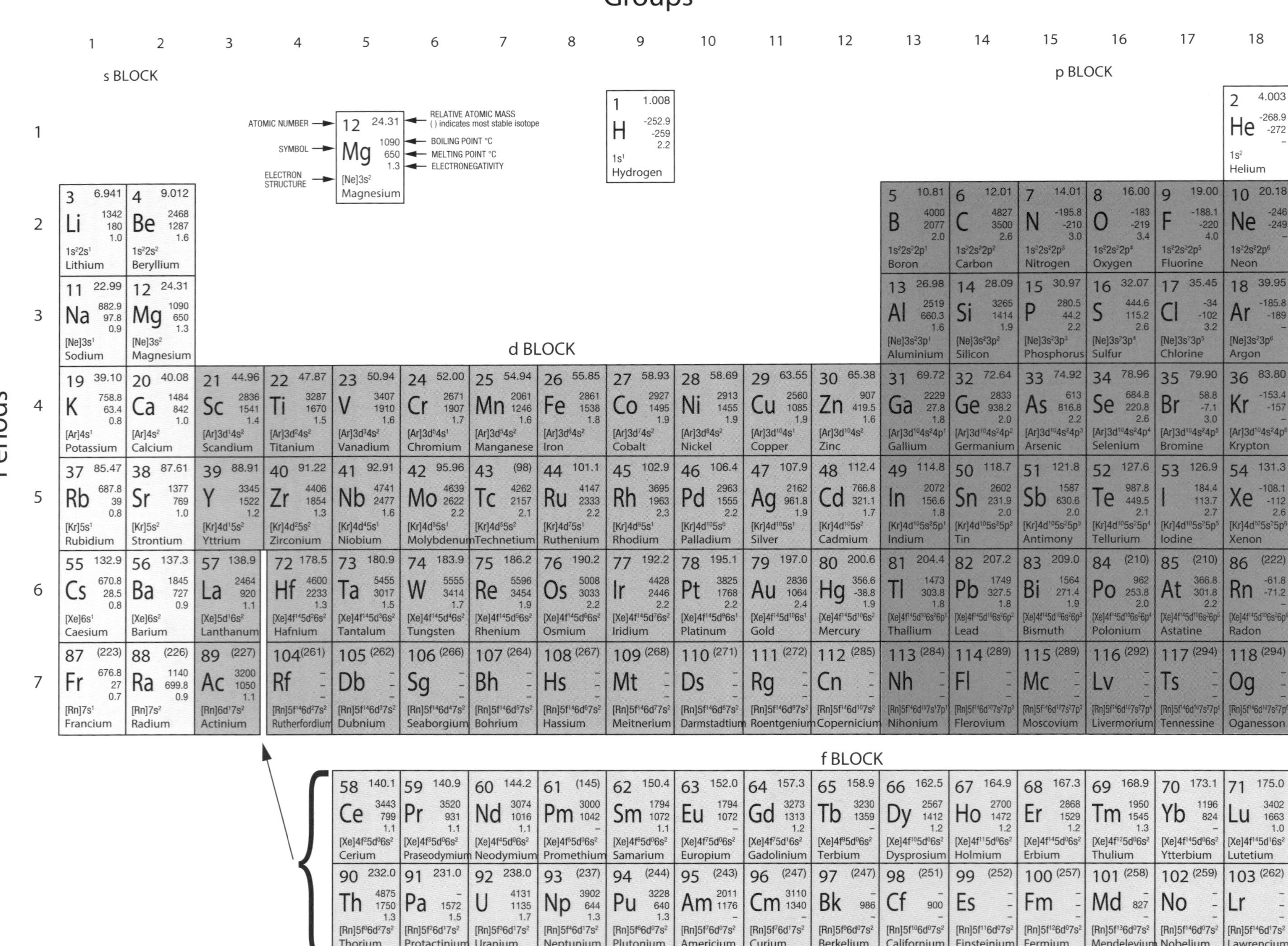

ISBN 978 1 4886 1933 5